How to Clone the Perfect Blonde

To Matthew Scott – the original Nelson-Hollingham production

How to Clone the Perfect Blonde

Making Fantasies Come True with Cutting-edge Science

Sue Nelson and Richard Hollingham

EBURY
PRESS

1 3 5 7 9 10 8 6 4 2

First published 2003 by Ebury Press,
An imprint of Random House,
20 Vauxhall Bridge Road, London SW1V 2SA

Random House Australia (Pty) Limited
20 Alfred Street, Milsons Point, Sydney,
New South Wales 2061, Australia

Random House New Zealand Limited
18 Poland Road, Glenfield, Auckland 10, New Zealand

Random House South Africa (Pty) Limited
Endulini, 5a Jubilee Road, Parktown 2193, South Africa

The Random House Group Limited Reg. No. 954009

www.randomhouse.co.uk

Printed and bound in Great Britain by Mackays of Chatham plc, Kent

A CIP catalogue record for this book is available from the British Library.

Cover designed by Keenan
Interior by seagulls

ISBN 0091892287

Contents

Acknowledgements

How to Clone the Perfect Blonde is a great concept (and title) for a book. Unfortunately it was not our idea. Otherwise we would have asked for a lot more money. Instead we must thank Andrew Goodfellow – the publishing world's equivalent of Agent Smith. Having said that, we came up with the contents and that, as any writer will tell you, is the hardest part. It has taken blood, sweat and tears to complete but only because one of us trapped a finger in the computer keyboard.

Naturally we would like to take all of the credit for the contents. Unfortunately, once again, we can't. The following scientists were of enormous help, ensuring that our own particular explanation of each scientific subject remained accurate. A few even suggested jokes – what a team. They are Harry Griffin, Derek Raine, Andrew King, Sam Braunstein, Jim Al-Kalili, Michael Heather, Paul Trayhurn, Johnjoe Mcfadden and Ray Mathias. It's at this stage that authors usually say they will accept responsibility for any mistakes. Fat chance.

Thanks also to Betsy Dresser, Robert Lanza, Arkady Tseytlin, Paul Hoiland, Malcolm MacCallum, Paul Edwards, Oliver Ryder, David Harris, Glenn Carter, Jo Webber, Nick Steneck, Elena Seymenliyska, Fred Adams and all the staff at the *Sunday Times* Wine Club. We are also greatly indebted to the BBC for allowing the extended career break necessary to write this book, together with the University of Michigan, whose journalism fellowship at Wallace House in Ann Arbor kick-started that break in the first place.

Writing a book with a toddler isn't easy. Probably because a toddler can't operate a computer yet and kept asking us to read a story. Friends and relatives offered permanent encouragement especially Peter, Colin, Melanie, Kate and Sarah. But for kindness, support and exemplary babysitting abilities, our greatest thanks go to Penny and Peter Hollingham and Bill Nelson. We expect they now know every Thomas the Tank Engine book by heart. As do we.

Preface

The idea behind *How to Clone the Perfect Blonde* is to take eight everyday fantasies and use each of them as a basis to explain big science ideas in an accessible, understandable and – naturally – best-selling kind of way. It has, we'll admit, made us think even harder about explaining science. Since both of us are science journalists, we are well aware of the enormous number of titles available in the popular science section of bookshops. This is aimed at those who couldn't get past chapter two of *A Brief History of Time.* Don't be ashamed, neither could we. So you'll be pleased to hear that the main qualification for understanding this book is not a PhD but an enquiring mind, an appreciation of science and a sense of humour. Which reminds us, if you follow any of the book's instructions, we accept no responsibility for the outcome. Or indeed anything.

The topics covered in each chapter – from cloning to the nature of consciousness – are at the cutting edge of science. As each topic could easily be the subject of an entire book, *How to Clone the Perfect Blonde* will enable you to gain both a wide and in-depth appreciation of these concepts. This is, we hope, an enjoyable read but it is definitely not science-lite. Certainly there's enough scientific detail to impress your mates and allow you to hold your own at intellectual dinner parties. And if any of the chapters encourages you to read another science book on the subject, then we've succeeded. If not, blame our Editor, it was his idea.

CHAPTER ONE

How to Clone the Perfect Blonde

Looking for love and affection? Know exactly what you want but haven't a clue how to get it? Relax. Science has the answer. All you have to do is clone the perfect blonde. Or brunette. Or redhead. You decide. With all the developing technologies of the twenty-first century at your disposal, think of the happiness cloning could bring. Provided you read the small print.

In a world where mankind's collective brainpower has invented computers, space travel and the self-cooling beer can, it makes sense to use another scientific advance for an equally life-enhancing experience. So whether your idea of perfection is a blonde or brunette, male or female, cat or dog, by understanding how to clone you could be one step ahead in that search for the perfect lifelong companion.

If playing God with genetics seems a radical way to avoid the angst of the dating game, it's time to wake up, smell the decaffeinated coffee and consider the less scientific alternatives: a course of evening classes or – if you're prone to gambling – a blind date. Even dating agencies can be too risky. Some may claim to have a scientific approach yet most rely on statistics. Tick the box and find a match. Hardly rocket science. And when that slim, blonde, blue-eyed Mensa member with a good sense of humour does reply, there's no guarantee this perfect match won't turn out to be a contact-lens-wearing brunette who lied about the IQ (don't we all) and will go to fat in five years' time (ditto). Imagine, then, if you could specify hair or eye colour, intelligence *and* body type by genetically selecting your ideal companion. Cloning could make this possible.

Still nervous? Then first some reassurance. Cloning, it has to be admitted, has one or two negative connotations. Mad scientists attempting to alter the world's natural order, the creation of cloned armies, babies on a conveyor belt – that sort of thing. The root cause of many of these fears is often science fiction. When genetic technology isn't being abused for personal gain (see origin of imperial storm troopers in *Star Wars* Episode II), it's being used to produce miniature Hitlers in the laboratory equivalent of a grow bag (see *The Boys From Brazil*).

However, as everyone except conspiracy theorists will concede, Hitler is dead. It's an important point to mention, because clones can only be made from a living cell. Cloning another führer is therefore a technical impossibility. More importantly, even if, for argument's sake, Hitler had faked his own death and is living in a rest home in Argentina, there is no guarantee that his clone would grow up to be a fascist warmongering vegetarian. Instead, the Hitler mini-me, through any number of life-altering events, may well be a meat-loving pacifist with an aversion to facial hair. Why? Because the environment we grow up in plays an important role in why people behave the way they do.

It is difficult to examine the full effect of environment when so many of us, despite living very different lives in different parts of the country, all watch the same television programmes, read the same newspapers and listen to the same radio stations. Yet there have been revealing studies of twins who had been separated through adoption or fostering. Some twins, even after living apart with no knowledge of the other's existence, were found to have many common events in their lives that suggested their shared genes affected the way they behaved. Other studies found that adopted children whose biological fathers were criminals are more likely to become criminals themselves, despite not living with the criminal parent.

So genetic inheritance is in some ways the key to who we are, but how we are brought up also plays a significant part in shaping our lives and behaviour.

Scientists are now beginning to incorporate environment in genetic studies. It is possible to witness how environment affects people simply by watching parents bringing up children. On the whole, a loving, caring, nurturing environment produces a loving, caring well-balanced child – at least until the teenage years kick in. Childcare books promote certain types of behaviour and non-violent forms of discipline for exactly this reason. If you rear a child in an atmosphere of fear and violence it can have a negative affect on that child for life, even influencing the types of relationships he or she enters into as an adult.

Interestingly, violence is a form of social behaviour that straddles both our genes and our environment. Research has linked violence with genetic components, nicknamed the aggression genes. This finding led one lawyer in the United States to mount a genetic defence for his client, Stephen Mobley, in 1991. Mobley was sentenced to death for shooting a pizza store manager in Georgia. As the murderer's family had a history of genetic illnesses and criminal behaviour, his lawyer claimed that Mobley was 'born to kill'. The appeal failed and Mobley remains on death row.

In 2002, a study published in the journal *Science* looked at a particular gene associated with aggression for a thousand children over a period of 30 years but also took the environment into account. It found that children whose genes produced low, rather than high, levels of a certain enzyme (monoamine oxidize A, or MAOA for short) were most likely to be antisocial in later life, but only if they had been maltreated or abused. This crucial finding proved an association between genes and environment which, if you think about it, makes perfect sense.

There is no avoiding the fact that nature and nurture both mould our lives. Scientists agree that it is a complicated mix and no one knows all the answers. What it does mean is that even if a Hitler clone could be successfully grown, the way the clone was brought up could easily upset that delicate recipe for a dictator. This is something *The Boys From Brazil*'s author, Ira Levin, also considered, suggesting that the teenage Hitler clone created by Nazi doctor Josef Mengele had to have a similar upbringing to the original Hitler for the

3

Cloning sci-fi style

To a younger generation (cough) *Jurassic Park* may seem like the first use of cloning in popular science fiction but in fact writers successfully hijacked the idea from scientists more than half a century earlier. In Aldous Huxley's *Brave New World* (1932) all humans are born using 'Bokanovsky's Process' (cloning) at the Central London Hatchery and Conditioning Centre. Society is divided into five classes from birth – alpha, beta, gamma, delta and epsilon – with specific classes being conditioned by a number of methods, from environment to electric shock treatment. Cloning here represents the ultimate control of human society and the complete lack of free will for individuals. There are plenty of other excellent novels dealing with cloning, from A. E. Van Voght's *The World of Null-A* to Arthur C. Clarke's *Imperial Earth*. Yet none, it is fair to say, have had the impact of *Brave New World*.

Now that computer software is sophisticated enough to produce almost any special effect, Hollywood has taken the lead in realising the subject's entertainment potential and no longer has to resort to look-alikes or dodgy editing to represent a clone. Even cloned armies can be convincingly created on screen, as seen in the Star Wars film, *Attack of the Clones*. Who would have guessed that the imperial storm troopers were all clones of bounty hunter Jango Fett? Or that the clone factories on planet Kamino could ever look so realistic?

Notice how the clones in both *Star Wars* and *Brave New World* are exploited, for military use or for society's benefit. Both types of clone are targeted for specific purposes. The Jango Fett clones, for instance, were also genetically modified to be better at taking orders. Neither type of clone has free will. This is a common representation of clones in fiction. Somehow their feelings don't matter, the implication being that they may look like the original but they are not the same. Instead, they are inferior copies bred for a certain task.

The prize for most spine-chilling representation of human cloning must go, in our opinion, to *Invasion of the Bodysnatchers*. Adapted from a great book by Jack Finney into a classic film in 1956, it is one of the best science-fiction films around, even though it has next to no special effects and the extraterrestrials are never seen. Somehow human beings are cloned while they are sleeping by giant alien plant pods. Physically identical to humans, these clones act as hosts for the aliens and, as the takeover continues, coldly, casually and without any regard for the originals take their place on Earth. The film was intended as a comment on communism and McCarthyism, but it also successfully undermines an assumption of what being human is all about: that we are each unique.

cloning to be a success. The same argument applies to anyone else's clone because a clone will be a clone not a copy. And while we're on the subject of what is or isn't possible, let's get on to the subject of *Jurassic Park* right now.

Michael Crichton's *Jurassic Park* is a great book and it made an equally entertaining film. Its premise was simple. Extract 'dino DNA' from dinosaur blood in the gut of a mosquito preserved in amber, combine with amphibian DNA (the only type of animal known to regenerate its own limbs), add a pinch of salt, heat at gas mark 4 *et viola*! One re-created cloned living breathing dinosaur. Can it be done? In a word, no, because unless a living dinosaur cell is discovered, the likes of T Rex and other 1970s rock bands will never roam the Earth again. Sorry.

SEND IN THE CLONES

People create clones – and eat clones – on a massive scale and on a daily basis. The person sitting next to you may even be a clone because, in 2000, more than 6,000 human clones were born in the United Kingdom alone. These clones have another, more familiar, name: identical twins.

This seems as good a time as any for a quick refresher course on sexual reproduction. For the whole process to begin, a male sperm must be introduced to a female egg. Assuming that no scientific assistance is needed, this will involve more than a handshake (no innuendo intended). Anyway, we're not going into the nitty-gritty of exactly *how* sperm is introduced into an egg except to say that if the two eventually get along and share the same interests, they fuse together. When the sperm penetrates the egg, fertilisation takes place and that wonderful thing called life begins. The fertilised egg cell begins multiplying into two, four, eight, sixteen cells, and so on to form an embryo. Nine months later, friends and family gather round a crib to decide whom the baby looks like.

Scientists have yet to explain why most newborns resemble Winston Churchill but, genetically speaking, the baby is a genuine 50:50 mix of its parents' genes. Half the genes are from the woman's egg; the other half are supplied by the man's sperm (which, under a microscope, all look like Woody Allen in a swimming cap). These hereditary genes determine a number of the child's key characteristics, such as eye colour, size of ears and tendency to go bald, wear an overcoat and smoke a large cigar. All siblings have the same proportion of genes from their parents but the mix will be different each time – in the same way you can never quite match a pot of paint several years after redecorating. Others have likened human reproduction to shuffling genes like a deck of cards, with the resulting hand containing equal number of cards (or genes) from each parent.

Identical twins, however, are a perfect genetic match. They occur when a single fertilised embryo splits into two and each embryo develops separately. This happens naturally in about one in every 60 human births, although the odds are higher if twins run in the family. The reason, by the way, that more non-identical twins are born through *in vitro* fertilisation, or IVF (where fertilisation takes place artificially in a test tube – *vitro* is Latin for 'glass'), is that multiple eggs are fertilised to give better odds of implantation in the womb.

While identical twins are genetically identical, they are not always physically identical – they can vary in height and appearance. Identical twins don't share the same fingerprints either, since these are determined in the developing embryo, but they do share the exact same genetic material and this makes them clones.

The artificial creation of clones is also not as unusual as you might think. Whenever a gardener takes a cutting from a favourite fuchsia plant to grow a new one, for example, a clone has been created. This cloned fuchsia is genetically identical to the parent plant, as are any other plants produced from cuttings in the same way. Even the word 'clone' originates from the greenhouse. At the beginning of the last century professor Herbert Webber, a plant-breeder and member of the US Department of Agriculture, needed a word other than bud, graft, runner or cutting to describe the plant sections that are removed for transplantation. A Greek dictionary must have been close by, as he decided upon 'clon' from the Greek word *klon*, meaning 'twig'. The word 'clone' soon followed. In the 1950s the meaning of the word came to extend beyond plants to living creatures, but before that it was used exclusively to describe crops, such as strawberries, that reproduce asexually and generate plants that are genetically identical to the parent.

COOKING WITH CLONES

Strawberries, unlike most berries, have seeds on the outside, but they do not always use these seeds to reproduce. Instead, slender side-shoots, or runners, grow out from a parent strawberry plant to form buds. These buds become roots and these roots then grow into the soil and form new plants. Strawberries are one of nature's clones because they are from a single parent plant and are genetically identical to that parent. Something to ponder when pouring cream on those clones during Wimbledon perhaps. Or eating chips for that matter, as potatoes can also reproduce asexually, via tubers instead of runners. You don't even need the whole spud either. Miniature sterilised cuttings the size of a single cell can now be grown in petri dishes to make tissue cultures and, under suitable conditions, multiple stems and roots result. The advantage of this method is its speed: one bud from a potato plant can be cloned up to a million times a year. Quite a sex machine I'm sure you'll agree.

Cloned plants don't stop at strawberries or potatoes either. Nature's clones also include raspberries, cranberries, garlic, grapes, pineapples, sugar cane and asparagus. Suddenly a grocery list of everyday cloned items takes shape. Stop off at a garden centre on the way back from the shops and the cloned products continue with ferns, orchids, spider plants and geraniums. Some people will even fly to the other side of the world and go snorkelling to see clones, in the shape of the coral of Australia's Great Barrier Reef.

While a night in with a bowl of strawberries or a baked spud is undoubtedly some people's idea of a perfect date, we realise that most of you are looking for slightly more animated company. Not to worry, there are certain types of insects, snails and shrimps that clone themselves. Even the toothless armadillo is in on the act, as it regularly produces quadruplets where all four offspring are clones. So given that we are surrounded by an abundance of clones on a daily basis (armadillos excepted), what made a sheep called Dolly such a celebrity clone when her birth was announced in 1997? The answer lies in the unique way she was created. Not quite the perfect blonde, brunette or redhead yet, but at least a creature sophisticated enough to make eyes at you.

HELLO DOLLY

It's not immediately clear why scientists would choose to clone a sheep. After all, most sheep look the same anyway. The clue to *why* Dolly was born lies in *who* created her: scientists at the Roslin Institute in Edinburgh and the commercial biotechnology company PPL Therapeutics, based at the same site. The Roslin Institute is a government research centre for animal biotechnology, sponsored by the Biotechnology and Biological Sciences Research Council. It receives extra funding from a number of sources, including industry, so the justification for this research originated from agricultural science and the commercial possibilities cloning could offer.

Since Dolly's birth in 1996, the same technology has been successfully applied to all sorts of animals, including pigs, cows, mice and goats. But if you're wondering what a sheep has to do with cloning that perfect blonde, it's this: the method by which Dolly was created could also allow you to clone your ideal partner. Unfortunately, unlike cloning plants, it's technically not as simple as taking a cutting of Jennifer Lopez (J Lo) or Jude Law (J La) and tending with loving care. The Roslin Institute needed 277 nuclear transfers, 29 developing embryos and 13 ewes to create just one cloned sheep.

OK, if it's that difficult, how did they do it? The answer is by using a technique called cell nuclear transplantation or nuclear transfer. Nuclear, in this case, has nothing to do with radiation. It refers to the nucleus – the mission control of a cell which contains chromosomes made of DNA, the basic genetic material of most living organisms.

The transplantation or transfer part of the phrases 'cell nuclear transplantation' and 'nuclear transfer' means exactly that. A nucleus is transplanted or transferred from one cell to another – in this case to an unfertilised egg cell that has been previously emptied of its own genetic material. Although this technique only really entered the public's consciousness through numerous newspaper articles, radio and television programmes about Dolly, nuclear transfer itself isn't new.

SPAWNING FROGS

Hans Spemann was a German zoologist, embryologist and Nobel Prize winner for medicine. He first proposed nuclear transfer, or cloning, in 1938 to test the role the nucleus played in development. Spemann called it a 'fantastical experiment', even though he had already performed a basic form of cloning in the 1920s that, by anyone's reckoning, was also pretty fantastical. He used a hair from his baby son's head as a noose to squeeze a salamander embryo with two cells, splitting them into two separate embryos, each of which grew into a salamander. He had produced identical twins.

All embryo cells are of immense interest to biologists. When a human embryo starts to grow, each new cell contains the same genetic material and has the potential to become a cell for any part of the body, be it skin, nerve, blood, heart or tooth. At the time scientists believed that once this process was under way it was a point of no return. When a bone cell started to become a piece of bone, for example, it couldn't change its mind and decide it wanted

DNA

Once you know that DNA stands for deoxyribonucleic acid you can immediately understand why its full name was shortened. DNA – doesn't that sound better already? – is a chemical inside the nucleus of a cell that carries the genetic instructions for making a living organism.

The DNA molecule (a particle made from two or more atoms) has a unique structure. It resembles a twisted, spiralling rope ladder, with the sides of the ladder made up of linked phosphates and sugars while the rungs are always two of the following four chemical components: guanine, adenine, thymine and cytosine. These chemicals, also called nitrogen bases, are more commonly known by their initials G, A, T and C. This explains the title of the science fiction film *GATTACA*, set in a society based on genetic perfection. The bases are limited in which ones they can pair with – only adenine to thymine (A–T) or guanine to cytosine (G–C). A single one of these base pairs, together with a sugar (deoxyribose) and phosphate molecule, is collectively known as a nucleotide.

Many people today would probably recognise the shape of DNA if they saw it, though in the early 1950s scientists didn't have a clue what this complex molecule looked like. Rosalind Franklin, at King's College in London, made a crucial contribution by applying a technique called X-ray crystallography to determine the structure of DNA. She discovered that the sugar-phosphate supporting structures were on the outside of the molecule. James Watson and Francis Crick then built on this work at Cambridge University. In 1953 they announced the shape of the DNA molecule – the so-called double helix formed from two strands of DNA linked by twisted parallel rows of guanine, adenine, thymine and cytosine – an achievement for which Crick, Watson and their colleague Maurice Wilkins were awarded the Nobel Prize. Franklin, the sole woman involved, was shamefully omitted.

While we're on the subject of genetic material, a gene is the unit of genetic heredity passed down from parent to child and is a piece, or stretch, of DNA. Most genes contain the information for making a specific protein and these proteins have lots of different functions. They can work alone or, more usually, in groups. A chromosome is a chain of hundreds of thousands of genes. We have 46 chromosomes (23 pairs) – half from our biological mother and half from our biological father. Sheep have 54 chromosomes (27 pairs). These chromosomes can be found in every cell of our body apart from red blood cells. The full complement of genes in an organism is called its genome. The human genome has often been referred to as 'the book of life'. In this analogy the chromosomes are the chapters, the genes are the sentences and the nucleotides are the letters of the alphabet. (For more about the genome see 'How to Lose Your Love Handles' on page 109.)

to be a liver cell instead. To use the technical terms, it had differentiated or specialised. The cell had already been chemically programmed and there was no turning back.

Spemann died several years after proposing his 'fantastical experiment' and never saw his ideas turned into reality. That feat was accomplished in 1952 by two American scientists, Thomas J. King and Robert Briggs. They performed the first cloning by nuclear transfer with the relatively common spotted northern leopard frog (*Rana pipens*). Frogs were a popular experimental subject because of the relatively large size of their eggs – they can be seen with the naked eye – and the fact that their embryos develop outside the body in spawn. Nuclei from early frog embryo cells supplied the genetic material and, although the cloning was not a very efficient process, some frogs did result. But, no matter how hard they tried, King and Briggs couldn't get nuclear transfer to work using adult cells. They concluded, as did other scientists who repeated the work, that it couldn't be done.

In 1962, Dr John Gudron took this research an important stage further using another species of frog, the pond-dwelling spotted South African *Xenopus laevis*. This extremely vocal species is well studied by biologists. The female frog has been described as using an 'acoustic aphrodisiac' to attract a mate. But that's not important right now ... because Gudron transferred the nucleus from a frog's gut cell into an enucleated egg cell and the resulting new cell began dividing and developing into tadpoles. Here, 'enucleated' means that the cell's nucleus has been removed. We cannot use the word 'empty' because, strictly speaking, the cell isn't empty when the nucleus is taken out. There is other material floating about the cell.

Each resulting tadpole was a clone, genetically identical to all the others – *and* to the frog that provided the gut cell. Gudron had done the seemingly impossible and made clones from adult frog cells. There was only one flaw: none of the tadpoles survived long enough to grow into frogs. Each scientist who replicated the work had the same result. Tadpoles but no frogs. One reason suggested for this anomaly is that there are usually some immature sperm or egg cells migrating out of the lining of a frog's stomach. These would definitely cause the development of a frog and so it's possible that the only successes, the cloned tadpoles, were from these particular types of cells and not adult intestinal ones.

Although a limited success, the cloning of a frog is a long way from the cloning of a handsome prince. The next breakthrough was made by Steen Willadsen, a Danish scientist working at the Institute of Animal Physiology in Cambridge. He cloned a live lamb in 1984, using the cell nucleus from an early sheep embryo – the first time this had been done with a mammal. It didn't take long for people to realise that one of the benefits of this technology was that it could be used by laboratories to clone valuable cattle. Willadsen himself joined an American company to do just that a year later, and in the following five years so many sheep and calves were cloned from embryo cells that cattle cloning became commonplace.

Despite this commercial application, cloned embryos from prize-winning cattle only have a limited use. There is no guarantee, for example, that the embryo will grow into a prize-winning adult and prove its pedigree. Scientists realised that cloning from an adult cell would be far more useful because you would then get a clone of a cow or bull that is a known quantity.

Willadsen continued experimenting with older embryos – some containing up to 128 cells – and, surprisingly, despite these cells being partially specialised or differentiated, managed to clone live calves. This work was never published, but one man heard about it over a drink in a bar at a scientific meeting. He was inspired to believe that cloning from even older, adult cells might be possible. That man was Ian Wilmut.

Wilmut worked at the Roslin Institute in Edinburgh, and he and his colleague, Keith Campbell, were also pushing the boundaries with sheep cloning. In July 1995 two sheep, Megan and Morag, were born. They were the first sheep to be cloned from embryo cells that had been grown for several months in a laboratory at Roslin and had started to differentiate. Just one year later Wilmut and Campbell did the seemingly impossible: they cloned a mammal from an adult cell.

I SAID HELLO, DOLLY

On 5 July 1996 no one outside the Roslin Institute was aware that a celebrity sheep had been born. In fact, it was another seven months before, in February 1997, the birth was officially announced in the science journal *Nature*. It caused, it is fair to say, a sensation – although this is hard to believe considering the less than headline-grabbing title of the research paper: 'Viable Offspring Derived from Fetal and Adult Mammalian Cells'. Now you know why so few scientists enter journalism.

There's no mention of the word 'cloning' (only its scientific definition, 'cell nuclear transfer') – or 'Dolly' for that matter. Just lamb 6LL3 – the one live birth cloned from an adult cell. The scientists who brought the world lamb 6LL3 were Ian Wilmut, Keith Campbell, Angelika Schnieke, Jim McWhir and Alex Kind. In developmental terms, Dolly proved that scientists could now effectively turn back a cell's biological clock and start all over again. To the world at large there was something much more controversial and sinister surrounding the birth of this particular cloned sheep, one whose mother was her identical twin. Dolly's very existence opened up the technical – and, to many, frightening – possibility of cloning humans.

THE SCIENCE OF THE LAMBS

Although Dolly received all the attention, three sheep made her birth possible: one Finn Dorset, a white breed; and two Scottish Blackface sheep, which have black or black and white legs and, as their name suggests, faces. The cells that provided the genetic material were from the udder of a six-year-old Finn Dorset sheep, while the eggs belonged to a Scottish Blackface ewe. The surrogate sheep, the mother who brought Dolly to term, was also a Scottish Blackface. The different breeds would provide an immediately visible indication of success, as a Blackface ewe would never normally give birth to an all-white sheep.

Technician Bill Ritchie transferred the nuclei from the udder cells of a six-year-old white Finn Dorset sheep into enucleated eggs (the nucleus had been removed) from a Scottish Blackface. The Blackface had been previously treated with hormones to stimulate the release of extra eggs. The egg cells were unfertilised but had been removed (and emptied of their genetic material) not long after ovulation so that they were on the brink of development given the right stimulation. All the donor cells had been starved of nutrients so that they were in a state known as quiescence – which means that the cells were effectively resting instead of busy dividing – a condition Keith Campbell thought would be beneficial to the whole process.

In a way that won't be lost on Frankenstein fans, a spark of electricity is crucial to the whole procedure. Imagine, if you will, a sheep strapped on a metal table, giant condensers reaching skyward as sparks fly around the lab. A scientist pulls a massive switch so that electricity fuses the egg cell and nuclear material together and kick-starts his creation into life ... Actually they did it all in a small test tube, using extremely narrow glass tubes called micropipettes to hold the egg, sucking out the nucleus and injecting the replacement genetic material before flicking a tiny switch to produce a small electrical pulse. Oh yes, and throughout the procedure a strong microscope is used to see what is going on. But you get the picture. It wasn't as painstaking as it sounds either. Each nuclear transfer only takes between five and ten minutes.

True clones?

After Dolly's birth, questions arose over whether she was an exact clone from an adult cell. Scientists formally challenged her authenticity in a letter published in the journal *Science* in January 1998. There were suggestions of possible contamination and mislabelling of cells. It was also suggested that because the six-year-old ewe supplying Dolly's genetic material had been pregnant, this may have affected the experiment. The crucial cell may not have been an adult cell, for instance, but a stray fetal cell.

Two scientific papers in the journal *Nature* eventually confirmed Dolly's claim to fame, using several techniques, including DNA fingerprinting. They also ascertained that the chances of contamination were minuscule.

Some of the public doubts about her authenticity were due to how DNA is distributed in a cell. In sheep almost all of the cell's genetic material, around 99 per cent, is contained in the nucleus, but – as in all living creatures including humans – a small amount of genetic material is found in the egg. It is called mitochondrial DNA and is found in mitochondria – a cell's powerhouse, or battery, which breaks down food molecules to release energy for the cell to survive. Mitochondrial DNA is only found in an egg so the biological mother passes it down, from generation to generation.

Nuclear transfer therefore creates a nuclear or DNA clone. The genetic material of the clone is exactly the same as the genetic material of the donor nucleus. In humans the percentage of mitochondrial DNA is even less than in sheep: about 37 out of 30,000 or so genes. Any human clone made using nuclear transfer would therefore be a true nuclear clone that would supply 99.9 per cent of the clone's total DNA.

Yet it was not an easy achievement. Two hundred and seventy-seven fused cells were cultured in the uteri of sheep for five to six days until they developed into the early stages of an embryo, a ball of cells known as a blastocyst. Only 29 blastocysts developed normally and these were then implanted into 13 Scottish Blackface ewes. One live birth resulted: lamb 6LL3, a clone of the

Finn Dorset sheep whose mammary gland cells also provided the more recognisable name, Dolly – after the busty country and western singer Dolly Parton.

Dubious taste in names aside, it was a tremendous feat. Wilmut's team was the first to clone a mammal from the genetic material of an adult cell – not through embryo splitting, artificially dividing the embryo in the same way identical twins form naturally; nor, as in previous experiments, by using embryo cells days after fertilisation. They did it by using a differentiated 'been there done that' kind of cell. A mature adult cell that, at the time, scientists considered too set in its biological ways to get to grips with all this new-fangled technology. They had reprogrammed an adult cell that was thought incapable of change. Instead, rather like that unsaved document on screen when your computer crashes, the cell's slate was wiped clean and it effectively started life all over again. Dolly was the result.

But why clone a sheep? Well, there are a number of reasons why a sheep gained the distinction of being the first mammal cloned from an adult cell. Cost was a factor. Despite the price of lamb in supermarkets, sheep are cheap – about ten times cheaper than a cow – and when you are on a government research grant these things matter. But the real driving force behind Dolly's creation was medical. The scientists involved were looking at ways to reproduce animals that they could genetically modify to contain important human proteins in, for example, their milk. Any animal that has been genetically altered in this way is known as transgenic because it contains genes across species. The concept promises enormous benefits both medically and commercially, and has already met with some success.

IT's CLONING CATS AND DOGS

Nuclear transfer has successfully produced cloned mice, cattle, pigs, goats, rabbits and, in February 2002, a kitten. Called CC (short for carbon copy) but nicknamed Copycat by the press, CC was cloned using nuclear transfer from cumulus cells – the cells that surround a mammal's egg in the ovary. The first kitty clone was born to a tabby surrogate and, like her genetic mother

Rainbow, was a three-coloured tortoiseshell. DNA tests proved that CC was genetically identical to Rainbow, although not an exact physical copy. CC's coat pattern was subtly different because this feature is determined in the womb, not by the genes.

CC was the only surviving kitten from 87 implanted cloned embryos in eight cats by scientists at the College of Veterinary Medicine at Texas A&M University. Despite these odds, pet lovers around the world had cause for celebration. Imagine being able to clone your beloved cat before – or after – she died? The company who funded this achievement, Genetic Savings and Clone (yes really), had already envisaged this dream. They will collect tissue samples and store your pet's DNA in a liquid nitrogen gene bank until the technique becomes a commercial reality.

As we know, it takes all sorts and some people's idea of the perfect blonde is a golden labrador. Well take heart dog lovers, because a multi-million-dollar dog cloning project, Missiplicity, is under way, named after a dead mongrel called Missy (1 April 1987–6 July 2002, RIP). Missy's anonymous billionaire owner provided most of the funding but the dog double has yet to materialise, as canine cloning is proving much more difficult than cat cloning. The same method is being used – removing the nuclei from other dogs' egg cells and inserting DNA from one of Missy's stored skin cells – but a dog's reproductive cycle differs from that of other mammals and is notoriously unpredictable. Dogs' eggs mature before the dog is on heat, and that only takes place every few months. This makes it harder to collect eggs and to synchronise any implantation after fertilisation, because a dog has to be physically ready to accept a cloned embryo. Scientists tried solving this problem by freezing dog embryos so that they could be put on ice until a dog is in the mood, but the freezing itself destroys the embryos, as the large amounts of fat in the cells cause them to crack at low temperatures. Also, while cumulus cells were successfully used to clone CC, Missy had been spayed and so didn't have any ovaries. This may turn out to be an additional hindrance.

Assuming that Missy will eventually be cloned, there is one important fact that her former owner must remember. Cloning is not resurrection and while certain traits are genetically determined, Missy 2 will not necessarily have the same temperament as Missy 1. This applies to all potential pet clones. Or, as the project's director Mark Westhusin put it, 'They're not getting Fluffy back.'

Research at North Carolina State University and Texas A&M University confirms this opinion. In April 2003 two separate studies on cloned pig litters demonstrated that the behaviour of the clones was not identical. The physical differences were also apparent, with genetically identical pigs showing varied hair length, weight and sizes.

For animal lovers, this is probably a minor point and until the technique is perfected Genetic Savings and Clone is one of several companies that offer a gene banking service, convinced that it is only a matter of time before man's best cloned friend starts chasing a copycat. Just be prepared to flex some plastic, as gene banks don't come cheap.

THE IDAHO GEM

There are some who think the world can be divided into dog and cat people, but there are also those – especially among the royal family – who are definitely horsey types. Perhaps that's why the press was so excited about the birth of a cloned mule called Idaho Gem in May 2003.

Idaho Gem was the first member of the horse family to be cloned. Scientists from the University of Idaho used a horse egg and a cell, which contained the genetic material, from a mule fetus. It wasn't easy. Over four years there were 307 attempts and it was only after calcium levels were altered in the fluid surrounding the eggs during cloning that eventually three out of 21 pregnancies were carried to term.

A businessman financed the project and there are a number of potential spin-offs – financial and medical. Idaho Gem's genetic material is from champion racing stock and, as mules are usually sterile, cloning offers owners the only way to breed from proven race animals. Naturally, some sections of the horse industry are extremely excited by the progress as the researchers claim their method could successfully be used on horses too. But Professor Gordon Woods, one of those involved in the cloning, was keen to point out the medical significance: the cloning of a new animal model could advance the understanding of human cancer and other diseases.

NOAH AND HIS FROZEN ARK

Cloning mules and beloved pets is all very well, and isolated farmers may not want to look any further than the nearest sheep, but if we are going to graduate to a perfect blonde, brunette or redhead, how about trying something more exotic?

Four hundred years ago, dodos were first seen by Portuguese sailors on the island of Mauritius in the Indian Ocean. They were big and extremely odd-looking – the dodos, not the sailors – and couldn't fly, so perhaps it's not surprising they became extinct. All that now remains of the dodo is the expression 'dead as a dodo' and the odd skeleton in a museum case. In today's more environmentally aware climate few people would want to see the giant panda or the Sumatran tiger head the same way. If your desire for an animal companion therefore extends beyond Rover and Fluffy, cloning could help – assuming you can find a big enough collar for the tiger and know a bamboo wholesale store for the panda.

Cloning is being seriously considered as an additional form of conservation to prevent endangered species from biting the evolutionary dust. There are plenty of reasons why. DNA can survive for about 10,000 years, so provided they have a good quality sample and the correct storage procedure, conservationists are in business.

For cloning to work, cells must be collected from either living animals or those that have recently died. The cells must then be frozen within five days and cryogenically stored. Luckily, and with amazing foresight, San Diego's Frozen Zoo has done exactly that for more than 25 years. Created in 1975 alongside the zoo's Center for Reproductive Species, it originally collected the animal cells for genetic studies during quarantine, routine health checks or after the animal had died. These cells, mostly from skin, are then cultured. Which sounds as if they were exposed to opera and classical music and taught deportment, but actually means they were grown in the lab in a suitable chemical solution. When ready they are frozen inside cryoprotected vials in liquid nitrogen at minus 196 degrees C. It was only later that scientists realised that there would be other uses for the cells they had collected.

Today the Frozen Zoo, led by geneticist Oliver Ryder, is one of the rarest repositories of its kind, containing more than 6,000 tissue samples from 400 different animal species, including pygmy marmosets, Malaysian tapirs and Siberian musk deer. And although Ryder is on record as saying cloning could never be done, the Frozen Zoo was in fact involved in the world's first cloning of an endangered species. This feat was reported, in January 2001, by Advanced Cell Technology (ACT) in Massachusetts. They cloned a gaur, a rare Asian ox, usually black or brown with white or yellow 'socks' on its legs and a ridge on its back.

Noah, the newborn gaur, was cloned using the genetic material from the frozen skin cells of a male gaur that had died around a decade beforehand at San Diego Zoo. The frozen gaur DNA was thawed and then fused with enucleated cows' eggs. It took 692 eggs to produce one live clone. Noah was born to Bessie, a domestic cow. Unfortunately, Noah died within 48 hours of birth from gas gangrene, a common bacterial infection in cattle that Ryder says was not related to the cloning process.

Ten months later, in October 2001, the University of Teramo in Italy completely big-footed ACT, announcing that they had successfully cloned an endangered European mouflon. No, we'd never heard of a mouflon either. It's one of the smallest sheep in the world, which is presumably why we'd missed it. By then, the cloned mouflon was seven months old and living in a wildlife centre in Sardinia, which meant the Italians had created the first viable clone of an endangered species.

Meanwhile ACT is working with other frozen San Diego animal cells to produce a banteng, a type of endangered wild cattle found in south-east Asia.

The real clone wars

Competition between different commercial companies – each producing their own cloned animals with rival cloning technologies – is heating up. These clone wars take place in court rather than outer space but they can be just as nasty. The three main players lined up for legal battles over cloning patents are Geron Corp, which now owns most of the rights to the Roslin patents; Advanced Cell Technology (ACT) and Infigen. All these biotechnology companies are American, and whoever wins could make millions. The winning company will be able to collect fees and royalties from whoever uses their particular cloning technique to produce, for example, cloned cattle, which are genetically engineered to produce valuable human proteins in milk.

At the moment technology is advancing so fast it's a bit like the wild west and there's a lot of confusion. The cause of most of the squabbling is over exactly how the nuclear transfer is performed. The Roslin technique for Dolly, employed by Geron, used quiescent cells – those that are resting and not dividing. Advanced Cell Technology successfully cloned the first cow using a slightly different method, with actively dividing – or non-quiescent – cells. They hold the intellectual property on that process. As a result, there has already been an unseemly public spat between the two companies, with Geron claiming that there must have been a stage when ACT's non-quiescent active cells were quiescent and resting. If that was the case, ACT would have infringed Geron's patented process. The claims have been refuted and so far no litigation has taken place.

Infigen also clones livestock and had a separate dispute with ACT. This revolved around patent interference concerning three animal patents, a case which ACT won in March 2003, confirming the 'value of ACT's intellectual property'. Another patent dispute between two cloning companies is expected to take years to resolve, and with so much money at stake and an increasing number of other companies entering the technological fray, the clone wars are set to continue.

The DNA, from stud number 319, belonged to a banteng that died at the zoo in 1980. It had been on ice for more than 20 years but, because of the conditions in which it had been kept, the DNA was in perfect condition. The banteng DNA was inserted into enucleated cow eggs, and a collaborating company, Trans Ova Genetics, transferred embryos into 30 Angus cows. Sixteen pregnancies resulted, with two live births at the beginning of April 2003 by caesarian section. One was a normal 20kg; the other was almost double the expected weight at 36kg. After feeding difficulties, the abnormally large calf was put to sleep a week later.

The giant panda, practically an emblematic image of endangered species, is proving a tougher nut to crack, but a successful cloning may well happen within the next few years. In June 2002 scientists from the Chinese Academy of Science announced they had cloned a panda embryo and implanted it into a host animal. They used the genetic material from a dead female panda and the egg cells of a white rabbit. The next stage is to be able to successfully grow the panda embryo in a surrogate womb. The most likely candidate as surrogate panda mum is either a rabbit, a cat or a bear, as these are genetically closely related to the panda. As you may have noticed, there's a slight difference in size between a giant panda and a cat or a rabbit. Fortunately for the potential surrogate mother, a baby panda is similar in size to cat and rabbit offspring. Otherwise, ouch.

While there have been some successes with closely related surrogates, not all species are genetically compatible. A successful cloning also depends on having near perfect DNA and a functional nucleus because then you get the whole genome and a complete set of chromosomes. Scientists in Australia have been trying to clone a Tasmanian tiger, or thylacine, since 1999, after finding a specimen of the animal – extinct since the 1930s – in a Sydney museum. There are several problems here – not least because there is no living animal closely related to the thylacine to act as a surrogate mother. The main stumbling block is the quality of the DNA ('hopelessly fragmented' as one scientist put it) since the animal was preserved in alcohol instead of by the more commonly used freezing process.

In May 2002 it appeared that this problem might have been resolved. Scientists at the Australian Museum reported that they had replicated some Tasmanian tiger genes using a technique called PCR (polymerase chain reaction). PCR has often been called a 'molecular photocopier'. This is because it makes use of polymerase enzymes (biological catalysts) that can help make millions of copies of DNA segments and build new DNA chains. The process takes advantage of the fact that we know that its bases (those four chemicals on the DNA ladder rung) must always be paired up adenine to thymine or guanine to cytosine. This is a relatively recent technology, devised in the early 1980s by biochemist Kary Mullis, and later earning him a Nobel Prize. It is the same method employed by forensic scientists in DNA fingerprinting at the scene of a crime.

When the Australian scientists used PCR, they discovered that the short fragments of Tasmanian tiger DNA were undamaged. This led them to believe there was no reason why the thylacine's genetic material shouldn't work in a living cell. Scientists from Oxford University who worked on DNA from an extinct giant bird, the New Zealand moa, are less optimistic. They concluded that any attempt to bring back extinct species using cloning would be futile. The main problem is that DNA deteriorates over time and has a lifespan of around 10,000 years, so ancient DNA is often damaged. Although PCR can replicate strands of DNA, it is still not good enough to re-create the entire genome of a prehistoric animal.

Similar problems lead to large amounts of rain falling on the likelihood of a cloned woolly mammoth parade. This despite the excitement caused when reindeer hunters found a 20,000-year-old mammoth, hair and all, buried in ice beneath the Siberian permafrost in 1997. The idea was to inject the mammoth's DNA into the egg of an Indian elephant, its closest living relative. So far no usable DNA has been extracted from a woolly mammoth. Unless you have a perfectly preserved piece of DNA the only way to clone a prehistoric animal is to find a live one. So until someone catches the Loch Ness monster, this will not happen unless we venture into science fiction once more and find The Land that Time Forgot.

As if the difficulty of obtaining a full set of chromosomes weren't enough, there are other stumbling blocks when it comes to maintaining a species through cloning. In 1999, Celia, the world's last remaining bucardo (a Spanish mountain goat), was killed by a falling tree. Obviously, not the luckiest of species. Fortunately, Celia the bucardo had provided plenty of tissue samples before her death. However, even if the cloning is successful, all of Celia's clones will be female and this means there will be no male bucardos to keep the species going. Possible solutions to this problem include either cross-breeding any bucardo clones with the male of a related species or combining their DNA to create a male bucardo.

No one has yet produced a live cloned primate by nuclear transfer, although the Oregon Regional Primate Research Center has cloned rhesus monkeys by embryo splitting, the way in which identical twins are formed. One thing is certain: if God decided to flood the world once more and wanted to save every species on Earth, the modern-day Noah would be a geneticist and his ark a frozen zoo.

CLONE SHOPPING

Assuming you now fully understand what goes into this whole cloning process, it's time for the part that will, in all likelihood, genuinely break the bank: cloning that perfect blonde. Or brunette. Or redhead. Like all projects, it's best to take things one step at a time and start with a shopping list. This is what you'll need:

- ❀ One human egg (check it's not past its sell-by date and look for the lion mark)
- ❀ One adult cell from your idea of perfection, preferably with their permission (must have DNA intact)
- ❀ A woman willing to give birth to the clone
- ❀ Test tubes (a range of sizes)
- ❀ Microscope (strength: strong)

❀ Pipettes (small ones)
❀ Electricity (check meter)
❀ One Micromanipulator (a geneticist's tool kit which is attached to your microscope and has a work surface and set of miniature tools delicate enough to carry a human egg without damaging it or letting it roll off the edge the moment your back is turned).
❀ One home laboratory. (If you don't have this, then getting a lab tooled up with the right equipment will set you back somewhere in the region of £100,000. If that means selling your home, consider hiring a lab instead.)

If you are male, then hopefully you will have noticed that something is missing from that shopping list – and, in fact, from the whole cloning process. There's no mention of sperm. The process is, to coin a phrase, spermlite – a realisation that can make some men nervous. The reason is simple: unlike normal reproduction, cloning does not need sperm. The genetic material can be obtained from a cell. Fortunately, there's always a need for someone to open a tightly closed jar, so I wouldn't worry too much.

Back to that shopping list. One human 'host' egg. Sounds easy doesn't it? But if cloning with sheep is anything to go by, hundreds of human eggs may well be needed to produce just one live birth. It must be done using a fine needle to collect the fluid containing the eggs while the woman is sedated either with painkillers or an anaesthetic. Between five and 15 eggs can be collected within about 40 minutes, and it is, reportedly, pretty stressful for the women concerned. Not surprisingly, human eggs are in short supply. IVF clinics, however, have eggs to spare, as part of the IVF procedure involves giving female patients hormones to stimulate egg production. However, in the UK these eggs are not for sale and their use is strictly regulated by the Human Fertilisation and Embryology Authority (HFEA) so it's more than likely you will need a volunteer.

Getting hold of an adult cell is much easier, but as this contains the valuable source of genetic material be careful with your selection: it determines what your clone will be like. If you've got permission from your clone of choice, then it's just a case of either getting a mouth swab or some other form of cell tissue. Contrary to popular belief, while DNA analysis from a strand of hair can help forensic scientists identify a crime suspect, the hair itself can't be used for cloning, because the hair on your head is no longer living. The same goes for dandruff, as those white flakes are dead skin cells not living ones. It's the same story with nails too. Remember, for cloning to be a success you need the genetic material from a living cell (or one that has been cryopreserved). If the hair happens to be attached to a follicle then that's different, as there will be a living cell containing DNA at the end of the hair.

If you haven't got permission from your chosen celebrity clone, say Pamela Anderson or George Clooney, getting that cell sample will be more difficult. (It might be also worth doing some research to see if your celebrity is naturally blonde in the first place.) Even if you get past the velvet rope and those scary bouncers, it's unlikely that your celebrity will provide a mouth swab instead of an autograph. This leaves the antisocial accidental bump and scratch routine, where you either leave in the arms of one of those bouncers or with some living celebrity cells under your fingernails. No one said this would be pleasant.

Thankfully, getting hold of the necessary equipment on that shopping list will be much easier and, assuming you remembered to use your scientific supermarket reward card, we're ready for business.

HUMAN CLONING

When the cloning company Clonaid announced – without proof – that it had cloned the first human being on 26 December 2002, it was case of *déjà vu*. More than 20 years earlier someone else had made exactly the same claim.

On that occasion, in 1978, the 'human clone' scoop was made by a respected science journalist, David Rorvik, in the book *In His Image: The Cloning of Man*. It was, at the time, a far-fetched story and, naturally, the book was a best-seller on both sides of the Atlantic. In it Rorvik detailed how he had secretly helped a millionaire who wanted to clone himself. This was supposedly done using nuclear transfer, with Rorvik helping to recruit the personnel in return for writing his book. Rorvik says he was never given all the details or proof of the cloning process's success, only circumstantial evidence, but he believed that it had happened. Scientists declared the whole thing a hoax for a number of reasons, lack of proof being one, but also because mammal cloning was then impossible. Nevertheless, there were some biologists who believed Rorvik because he cited the work of J. Derek Bromhall, a respected embryologist at the University of Oxford. Bromhall sued for libel and the lawsuit was eventually settled with the book's publishers admitting the book was untrue and paying Bromhall damages.

Today, of course, mammal cloning is possible, and after Dolly's birth there is no biological reason why we can't clone humans. The technology to do so has already been proved with another mammal using an adult cell. The main reasons holding most people back are ethical ones and those related to how it would be done. It is undeniably premature to view cell nuclear transfer as an efficient route for human cloning. The failure rate is high. For Dolly the chances were less than half of one per cent.

Success rates have undoubtedly improved since then. In 1998 Ryuzo Yanagimachi, Toni Perry and Teruhiko Wakayama from the University of Hawaii reported that they had cloned three generations of mice from adult cells. The process had worked for about three in every 100 attempts. The Honolulu Technique, as it is now known, is similar to the Dolly method but differs when it comes to the fusion of cells. Instead of jump-starting fusion with a jolt of electricity, the scientists let the egg, with its newly inserted genetic material, sit in a chemical bath for about six hours before the cells started dividing.

Although many people have pointed out that the mice are genetically more similar to humans than sheep are, the scientists involved did not intend these techniques to be applied to human cloning. But even if the technique were used, dozens of eggs would still be required to produce a live birth. Many women would have to be willing to undergo pregnancies with the strong expectation of miscarriage, stillbirth or birth defects. This is because the process can produce spontaneous abortions, physical deformities and unusually large offspring. Caesarean sections often have to be used to deliver cloned young because of the clone's birth weight and size. No one yet understands

Goodbye Dolly

On 14 February 2003 the world's most photographed sheep died. Dolly was six years old, a mother of six and a celebrity. Famous for being the first mammal cloned from an adult cell, Dolly led a pampered existence and received high-tech veterinary care. She was put to sleep after it was found that she was suffering from a serious lung disease.

A post-mortem found that she had Sheep Pulmonary Adenomatosis (SPA), a lung tumour not uncommon in older sheep. Sheep can live until they are 11 or 12 years old so Dolly was comparatively young when she died, although it is not unusual for SPA to affect sheep aged four or five upwards. SPA is caused by a virus, so it may never be known if Dolly simply caught this virus naturally or was at increased risk of catching the virus because she was a clone. Nevertheless, Dolly was no ordinary sheep and the fact that she only survived for half of her expected lifespan means that questions about her possible premature death need to be answered.

The Roslin Institute held a wake for their famous inmate. They said they believed that Dolly's death was unrelated to the cloning process and promised to make the post-mortem findings public. Roslin has always voiced its opposition to human cloning, and for them and others the implications of Dolly's early demise are too important to ignore. While there is still much to learn about the full effects of cloning, anyone who wants to produce human clones would be putting lives at risk. Yet just days after Dolly's death the fertility expert Severino Antinori pledged to continue his plans to clone humans.

The University of Edinburgh performed a detailed analysis of Dolly's tissues and cells, but at the time of going to print the results are yet to be published. Wilmut announced that there were no plans to re-clone Dolly, and she is now an exhibit at the Edinburgh's National Museum of Scotland.

why this happens. Some studies on cloned mice have also reported obesity and early death, and the long-term health of cloned animals is still uncertain.

In 2002 scientists reported that 11 out of 14 genetically modified and cloned lambs had abnormal kidneys, brain or liver and died within 12 weeks of being born. It remains uncertain whether Dolly's onset of arthritis, for instance, was premature and resulted from being cloned from an adult cell that was six years old. Did this make six-year-old Dolly's cells 12 years old at the time of her death?

Telomeres, the protective structures on the end of chromosomes, are often referred to as the cell's virtual ageing clock. When researchers in the US bred mice with shorter telomeres the animals went prematurely grey and died early. In 1999 the Roslin scientists reported that Dolly's telomeres were shortened by about 20 per cent. Many people wondered whether this meant she was genetically older than her physical age. The question has not been answered because measuring the lengths of telomeres is not an exact science and this 20 per cent difference could be within normal limits. Telomere length may not be an accurate prediction of lifespan either as cloned cattle have been found to have normal-sized telomeres. Nevertheless, the possibility of premature ageing renewed fears about the cloning process when Dolly's death was unexpectedly announced in February 2003.

Dolly wasn't the only famous sheep to die early in February 2003. So did Matilda. Not so well known here, admittedly, Matilda was big down under because she was Australia's first cloned sheep. Matilda's death was totally unexpected because she was only two years old and earlier that day had appeared perfectly healthy. The South Australia Research Institute said death was by natural causes after an independent autopsy failed to find out why Matilda, a merino ewe, had failed to reach her third birthday.

The carcass was cremated – prompting criticism among those opposed to cloning and genetic technology, as further tests could not then be done to verify exactly why Matilda had died so young and if the cause was related to the procedure that created her. Even so, the premature deaths of Matilda and

Dolly should act as a warning bell to anyone who is considering cloning a human being.

A few months after Dolly's death, in April 2003, a team of scientists from America added to the debate. A report in the journal *Science* stated that the failure to clone a primate could signal the impossibility of cloning humans. The researchers tried to clone macaque monkeys and found that, despite more than 700 attempts, no pregnancies had occurred and that there were chromosomal problems in each individual cell. Dr Gerald Schatten, the team leader, said: 'I don't want to say that this will never happen. Given enough time and materials, we may discover how to make it work. It just doesn't work now.'

THE CLONERS

The risks of human cloning have been highlighted, debated and highly publicised in almost every media yet there remain plenty of women prepared to give birth to a clone. Provided these women have the funds, there are scientists equally prepared to help them do so. The most prominent pro-human-cloning scientists are Drs Severino Antinori, Panayiotis Zavos and Richard Seed (no, that's not made up).

Antinori, an Italian embryologist, is the director of the International Associated Research Institute for Human Reproduction Infertility Unit in Rome and is well known for his IVF work, especially after he helped a 63-year-old woman to produce a child in 1994. He claims to have 600 couples on a cloning waiting list

Zavos is a fertility expert who once worked with Antinori but left to set up his own cloning team. He wears a number of hats – as a University of Kentucky professor, as director of his own Andrology Institute of America and as CEO of a private company that markets infertility products and technologies. He also founded Sperm R Us, a firm offering home semen analysis. An essential service for today's busy executive.

Seed is considered more of a maverick and unlikely to achieve his aims. He is a physicist with fertility experience who has said that cloning would bring us

The UFO connection

On 13 December 1973 an alien apparently visited our planet in a spaceship and introduced himself to French sports journalist and former racing driver Claude Vorilhon. As you do. Luckily for Vorilhon, it wasn't the monstrous 'Mmmm, humans taste good!' kind of encounter. This humanoid extraterrestrial was a four-foot cutie with green-tinged skin, almond-shaped eyes, dark hair and a good sense of humour. His name was Yahweh Elohim. During a series of hour-long morning meetings over the next six days, Elohim dictated his messages for humanity, revealed mankind's true origins and told Vorilhon to change his name to Rael, which means 'messenger'. You have to admit, it does have a nice ring to it.

For those who know their scriptures, the word 'Elohim' appears in the Bible, where, according to Yahweh Elohim, it is mistranslated as 'God' rather than 'those who came from the sky'. Worse still (from Yahweh Elohim's point of view), God took the credit for Adam and Eve when it was the Elohim – 25,000 years more scientifically advanced than us – who created life on Earth using DNA and genetic engineering. All the major religions were founded by Raelian ambassadors while Jesus' resurrection and reappearance outside the garden of Gesthemane could be explained scientifically: Jesus had been cloned.

The Raelian Movement now has more than 55,000 members worldwide, with most people probably having become aware of its existence in 1997. Immediately after scientists announced Dolly the sheep's birth, Rael founded Clonaid and openly declared it the world's first human cloning company. Now led by biochemist and Raelian bishop Dr Brigitte Boisselier, Clonaid attracts worldwide attention and condemnation for its aims but says it has enough funding and a sufficient number of volunteers willing to undergo human cloning experiments.

Clonaid claimed they had created the first human clone on Boxing Day 2002 – a baby called Eve, allegedly born by caesarian section to a 31-year-old American woman – but offered no scientific proof of this feat. Their ultimate aim is eventually to transfer memory and personality and obtain eternal life inside a computer (for more, see Chapter 8). If life is too short, go and visit UFOland in Canada, it's a theme park opened by the Raelians in 1997.

closer to God. Then there's Clonaid's Dr Brigitte Boiselier, a French research chemist and a Raelian bishop who believes space aliens created life on Earth.

These are the main scientists who are publicly claiming to attempt human cloning. There are doubtless others working in countries where human cloning remains legal and who prefer, for obvious reasons, to remain anonymous. The legal and ethical issues, along with most volunteers' desire to keep their actions out of the press, naturally results in a considerable amount of secrecy surrounding private attempts at human cloning. This makes verification of human clone claims extremely difficult.

In March 2003 Antinori said he had proof that human cloning worked but he did not provide it to journalists. One of his human cloned babies was also supposed to have been born several months earlier, yet no evidence was produced to back his claims. Similarly, Clonaid, after inviting a journalist to join an independent panel to verify their 'first human clone' headlines, have not shown the world baby Eve. The parents, they say, withdrew media co-operation after the birth. At the time of going to press, Clonaid claims to have successfully produced five human clones. Pictures of them can be seen on their website but, again, none of these babies have been scientifically shown to be clones.

From a scientist's point of view, the way Clonaid chose to reveal the news of their human clone is equally suspect. It was done through a press conference instead of the accepted procedure of publishing in a scientific journal. This is an important rite of passage within the scientific community for any piece of work and acts as a form of quality control. All claims, procedures and results are peer-reviewed by other experts in the field. That Clonaid chose to go to the press first makes scientists understandably suspicious.

UFO believers and would-be cloners aren't the only ones to attract the attention of the world's press. In 1993 two scientists were the unwelcome targets of human clone headlines when Robert Stillman and Jerry Hall (not that one) performed the first artificial twinning or splitting of human embryos – nature's way of cloning – at George Washington University in the US. They used the same procedure that had already proved successful in twinning mice and cattle embryos, using fertilised eggs. One aspect was different.

A fertilised egg has a protective membrane called the zona pellucida. It contains the nutrients for the first few cell divisions. Stillman and Hall used a chemical solution to dissolve this coating and then create a new one around separated embryos. The embryos, if they had been left to develop, would then have grown into two genetically identical human beings – although, as it happens, the embryos in this particular experiment could not have done this because more than one sperm cell was used.

The research caused an ethical uproar, not least because the scientists believed that by experimenting on early embryos consisting merely of bundles of cells they had not created or destroyed life. This 'where does life begin?' controversy continues to rage among religious groups and ethicists today.

WHY CLONE?

Strange as it may seem, not everyone wants to clone humans for the simple reason of producing an ideal companion for the bedroom or cocktail lounge. There are all sorts of complex reasons for cloning and no one should under-estimate the genuine love and longing involved in wanting a biological child.

The UK's National Infertility Awareness Campaign says that one in six couples is affected by infertility. The causes are not always discovered but range from sperm defects and ovulation problems to reproductive damage caused by diseases or infections after abdominal surgery. IVF treatments are expensive and stressful, and there is no guarantee of success. A couple's chance of an IVF birth also varies from clinic to clinic. It can be as high as one in five or as low as one in 20. For the couple who have exhausted all possible avenues, cloning may be their only hope. Cloning gives an infertile – or gay – couple the theoretical possibility of conceiving a child whose genetic make-up is derived wholly from one of the parents instead of a child half of whose genetic make-up is from the sperm or egg of an anonymous donor. A second child for such a couple could then be cloned from the second parent.

Old Macdonald has a pharm

Growing animals for pharmaceuticals – or pharming – is a genuine commercial application of cloning. Human proteins can treat cystic fibrosis, the blood-clotting disorder haemophilia and certain types of emphysema (lung disease), but these proteins are not always easy to make. They often require a human donor and this makes the proteins limited in supply and expensive. If human proteins could be obtained more readily and cheaply from the milk of transgenic sheep, goats and cattle, then a vast number of human diseases could be treated. This is one of the reasons that a mammary cell was used to create Dolly, because if you can genetically modify this type of cell to produce a protein then you can see if this ability is passed on to the adult sheep.

The first cloned animal to be genetically engineered to contain a drug was a sheep called Polly in 1997, again courtesy of the Roslin Institute and PPL Therapeutics. Polly carries a human gene for the blood-clotting Factor IX that is used to treat haemophilia. Her existence resulted from earlier work, including the genetic engineering that produced a sheep called Tracy, born in 1990. Tracy's milk contained alpha-I-antitrypsin (AAT), which treats emphysema and cystic fibrosis.

Another avenue for pharming is the production of a transgenic bird that could carry human proteins or vaccines in its eggs. Genetically modified pigs are also considered a possible solution to the shortage of human organs for transplants. Replacement heart valves from pigs are already used in many human patients, and pig hearts are a similar size to a human one. The main problem is that the human immune system fights back: because of the proteins surrounding the organ, the body rejects the heart. Cloning allows the elimination of the genes causing this rejection.

Transplanting animal organs, tissues or cells into humans is called xeno-transplantation. There is a serious shortage of human organs and people die waiting for kidney transplants. There are plenty of pigs. The maths, then, is simple. The science is slightly more complex, but if a pig is genetically modified so that its cell tissues have an added human protein around the necessary organ, the hope is that its new human host would not reject the pig's kidney, heart or liver. The human transplant waiting list would cease to exist. The main problem that must be overcome is the possibility of zoonoses – diseases that can be transferred from animals to humans.

There are others who believe that the technology could help replace an only child who has died in an accident. In his book *The Second Creation* (written with Keith Campbell and Colin Tudge), Ian Wilmut describes fielding calls from bereaved parents who wanted him to clone their dead children. These cases are distressing and disturbing but understandable when viewed within the context of grief. Cloning, remember, is genetic copying not resurrection.

Apart from infertility, the other main reason for cloning is for medical purposes. Scientists have discovered that shortly after fertilisation the embryo contains cells that can become almost any cell in the human body. These are called stem cells and are often considered the mother of all cells because they turn, or differentiate, into specialised cells such as liver cells, heart cells, nerve cells and so on. At an early embryonic stage, however, stem cells have not yet differentiated, and so scientists believe that if properly directed they could replace damaged cells from almost anywhere in the human body. They could repair heart muscles, brain cells and spinal nerves, and could rectify diseases such as Parkinson's, cystic fibrosis and diabetes.

Stem cells have the potential to revolutionise medicine and our current concept of healthcare. The problem of rejection by the body's immune system could possibly be overcome by cloning yourself at the embryo stage to collect your own genetically compatible stem cells. This process – cloning from early human embryos for stem cells – is known as therapeutic cloning.

This is, not surprisingly, an ethical minefield, in which people have to weigh up the life-saving treatments offered by cloning against the destruction of embryos that the process involves. And then there is the question of where we define life as beginning. When the embryo is a ball of cells, at the blasto-cyst stage for instance, it does not have a nervous system.

Therapeutic cloning is not, however, the only way to obtain stem cells. Blood from a baby's umbilical cord contains stem cells, and in the United States it is becoming almost routine to offer parents the choice of freezing the cord shortly after birth. Several companies in the UK are now offering the same serv-ice privately. This is seen as an insurance against any illnesses the child might have in the future. The reasoning is that if the child later contracts a disease that destroys its body cells, then there is a source of stem cells immediately available to be grown into whatever replacement tissue is needed. All of this involves an assumption that doctors will be able to use this form of medicine at some stage in the future.

Stem cells can also be collected from adult cells. Blood stem cells, for instance, are found in bone marrow and can be separated from the blood and collected in a laboratory. Although not as versatile as embryonic stem cells, adult stem cells are showing greater promise than originally thought and their use could be considered an ethically acceptable alternative to cloning embryos.

Other, less convincing, reasons given by some companies wanting to clone, include using perfect tissue matches in cosmetic surgery and breast implants and thus decreasing the likelihood of infection and rejection. (Treating burns victims would seem like a far worthier justification.)

WHY NOT?

The emotional reasons for human cloning and the medical reasons for therapeutic cloning are convincing, so why are so many people opposed or even repulsed by the idea? I know you just want to clone the perfect blonde, but there are a number of important ethical dilemmas to consider first. Think of it this way. Some people make an informed decision about whether to eat meat, wear real fur, buy organic food or frequent a small locally owned shop rather than a supermarket chain. For the biggest decision of your life it pays to think it through first.

Safety has got to be one of the main concerns surrounding potential human cloning. The fact that a single successful cloning involves so many miscarriages, abortions and birth defects is in itself deeply offensive to many people, particularly religious groups. This prompts questions on the morality of cloning in the first place and accounts for why many pro-life groups are against therapeutic cloning, despite the potential medical benefits, as early embryos are destroyed in the process of collecting stem cells.

Others wonder about the psychological impact of being a clone. Imagine how a child would feel knowing that he or she is a replacement for a dead brother or sister? It also prompts questions about free will. If the deceased was a talented artist, would the clone feel obliged or even coerced by parents into following the same artistic direction? Genetic cloning does not, as we have discussed, ensure that the clone will be physically or temperamentally identical to the person it has been cloned from. So even if someone had been able to clone Marilyn Monroe before her untimely death, Mary, the cloned (dyed) blonde bombshell, might have lacked the looks and that special something to make her a star.

There's also the prospect of growing up as someone's vanity project and the issue of control underlying the cloning in the first place. Even if a human clone is born out of a gay or infertile couple's deep desire to have a child, the fact remains that this child could be the identical twin of its mother or father, assuming they supplied the genetic donor material. Going through the hormonal swings of puberty is bad enough. Being your mother's twin on top of acne and boyfriend trouble would surely do your head in.

THE SMALL PRINT

We will assume that, by this stage, you have given human cloning a lot of thought. You have not only followed our instructions but have also gathered the appropriately qualified scientists; personally resolved the legal, ethical and technical difficulties associated with cloning and bought a white lab coat. You have now successfully cloned the perfect blonde. Or brunette. Or redhead.

Your clone will probably not have the same temperament or personality as the original and, of course, will not necessarily like you either. There's just one more thing you ought to know. By the time your clone has grown up, say 20 or so years later, there's also going to be a huge age gap between you and your ideal companion. This may not matter if you are Rod Stewart or Cher, but for the rest of us it means a long wait. Considering all the advantages and disadvantages, is human cloning really worth it? You decide.

CHAPTER TWO

How to Build a Domestic Goddess

If cloning a companion doesn't appeal, how about the next best thing – building a servant. It's creative, there are few ethical concerns and just think how useful it could be. Imagine how your life would benefit from you owning your very own domestic goddess. No more cooking, cleaning or ironing. No more fetching drinks, washing socks or making your bed. Can't afford it? Don't worry, because after construction costs this particular domestic goddess won't need paying, won't answer back and will fold away neatly under the stairs when all its chores are completed. Congratulations. You are about to become the proud owner of a multi-purpose household robot.

Let's call it Nigella.

If scientific predictions of the last 50 years are to be believed, household robots should be a reality by now and there ought to be at least one Mark 5 Nigella (the non-gushing model) in every home. Or, if you prefer, a Mark 2 Jeeves (the non-balding male model). Sadly for the couch potatoes among us, science has completely failed to deliver. It's the twenty-first century, so why, when robots can build cars, assemble space stations and defuse bombs, can't they fetch a beer or shift a stubborn stain?

Let's be fair, computers can do some pretty impressive things. Just manipulating the letters on this page takes considerable processing power. Storing, changing, deleting. Each new word represents thousands of digital calculations. But a computer operates within strict logical rules. When you type a letter it

Why, robot?

A Czech playwright, Karel Câpek, came up with the word 'robot' for his play *Rossum's Universal Robots*. He derived it from the Czech term for forced labour or drudgery. First performed in 1921, the play features man-made biological (as opposed to mechanical) workers built in a vast factory to replace humans, Rossum's robots being cheaper and more efficient than the people they displaced. Of course, it all goes horribly wrong and the robots end up taking over the world. This being an eastern European play, there's also quite a lot about the failures of capitalism and the free market. Even so, it remains eminently readable because the themes of machine domination are just as relevant today.

Although the word robot is a twentieth-century invention, the idea of some sort of artificial human (usually mechanical) has been around for centuries. Leonardo Da Vinci even drew up plans for one in the sixteenth century (along with a space shuttle and a cappuccino machine), although he never built it (ditto). Later, Victorian engineers developed automata powered by clockwork that resembled humans (in so much as they had the right number of limbs and could sit staring into space for an indefinite amount of time).

The science of robotics owes its name to the science-fiction writer Isaac Asimov. Between 1940 and 1950 he wrote a series of stories involving various technological, moral and philosophical aspects of robotics and artificial intelligence. By the end of the decade these had been published as a collection *I, Robot*. In this book Asimov conceived the 'laws of robotics', imagining robots designed to obey humans and protect themselves. Overriding everything is a stipulation that robots must not injure humans or 'through inaction' allow humans to be harmed. Asimov later added another law about not damaging humanity as a whole.

If the laws are broken, the robots of his stories have a tendency to go mad, shut down or just confuse the hell out of everyone. Robots turning on their creators is a common theme in science fiction, so here's an important survival note for future reference: if you're ever up against a mad computer/robot and your life depends on it (see *Star Trek*, *The Prisoner*, *The Avengers* or, in fact, any 1960s sci-fi), simply ask it, 'Why?' At this point smoke will most likely start billowing out of the central console, it will begin wailing 'cannot compute' and will burst into flames.

sticks that letter on the screen; it doesn't have to understand what it's doing. The machine has no concept of its place in the world, no common sense, no intelligence. I HATE MY COMPUTER. See, it did nothing (although to be fair it'll probably crash later). A basic computer is fundamentally stupid. Housework, believe it or not, requires a brain.

Here's what we mean. If you are at home, take a look around the room. You know where everything's supposed to go; you know what looks clean and what doesn't. It wouldn't take long to tidy up. No, really it wouldn't. Imagine a robot in the same position. The robot needs a vast amount of information before it can clean the room. It needs to know what is or isn't rubbish (in case it feeds your prized David Beckham autograph into the waste disposal unit). It must be able to recognise dust (in case it strips the paint from the walls) and where items on the furniture should go (in case it squeezes Uncle John into a book-case). A nice shiny silver robot may appear state-of-the-art but it needs lots of information and all sorts of rules to function in a sensible way.

The man who has subsequently become known as 'the father of robotics', Joseph Engelberger, developed the first commercial robot in the late 1950s. Called the Unimate, it was a robotic arm designed for heavy industry – it could lift, weld, and carry out boring and repetitive tasks – and it didn't demand union recognition. Modern robots can do everything from coring apples to vacuuming floors. But no robot can do both. Most are designed to only carry out specific tasks. Attach a feather duster to an apple-coring robot and it will try to core an apple with it. So unless you want a Stepford Wife style slave, the crucial ingredient for an all-singing, all-pouting Nigella (Mark 5) is a brain. It needs some *artificial* intelligence.

BRINGING UP BABY

The aim of artificial intelligence, or AI, is to re-create in electronic form something that acts in the same way as a brain (human or otherwise). An artificial intelligence would learn, remember, adapt and even think. It could have emotions and a sense of self but be designed for a specific purpose. In the

Steven Spielberg film *AI*, for example, based on a short story by Brian Aldiss, scientists create a robot child who can love – as well as a robotic gigolo, played by Jude Law, who can love in an entirely different way. The film *Bladerunner*, based on Philip K. Dick's story 'Do Androids Dream of Electric Sheep', also contains humanoid robots, but these renegade replicants have evolved a new purpose: to survive a built-in sell-by date, no matter what the cost.

Whereas a conventional computer is limited by a strict set of instructions – its program – an 'intelligent' computer could adapt to changing circumstances. Not surprisingly this is an idea that has excited scientists for decades. In the fifties, when computers the size of a small planet (or large room anyway) were crunching through equations now calculable on pocket calculators, researchers were already contemplating the potential of AI.

One of the first people to put his thoughts on AI in writing was the British mathematician Alan Turing. He proposed designing a computer program that would simulate the mind of a computer-child. Researchers could then 'educate' this cyber-kid to achieve the intelligence of an adult. In his 1950 article in the philosophical journal *Mind*, he likens a child's brain to that of a notebook with lots of blank sheets. It was up to the programmers to fill those sheets.

The cyber-kid's education would involve a system of punishments and rewards. That's not to say that he imagined the machines having feelings, but rather if the computer received a 'punishment' signal it would be unlikely to do the same thing again (sadly, the same doesn't always apply to children). The spooky thing about these learning machines was that the programmer didn't quite know what was going on inside. But Turing never got as far as building his intelligent computer.

MIND IN THE MACHINE

In the same article in *Mind*, Turing proposed what he called an 'imitation game', which has since become known as the 'Turing test'. Judges in one room pose questions via a keyboard to computers and humans in another. From the responses the interrogators have to identify which answers come from the people and which from the machines. By Turing's criteria, a computer that could fool the judges into thinking it were human would pass the test and could be said to be truly intelligent.

By 1990 researchers had even more of an incentive to compete, when the American philanthropist Hugh Loebner announced that he would award a solid gold medal and $100,000 'for the first computer whose responses were indistinguishable from a human's'. Turing predicted that by the twenty-first century a computer would already have passed the test, but although some have got close, his dream remains unfulfilled. As he said himself, however, 'If a computer could think, how could we tell?'

CHINESE WHISPERS

Assume for a moment that a computer is developed that passes the Turing test. Can it really be said to possess intelligence?

One of the fiercest critics of AI research has been the philosopher John Searle with his Chinese Room Argument. Not who had the spring rolls as a starter (that's the Chinese restaurant argument) but the one about intelligent understanding. He imagines himself in a locked room with a set of rules and a load of cards with Chinese symbols. He has no knowledge of Chinese; to him they are just meaningless patterns. A story in Chinese is passed through a slot in the door, along with some questions about it. By following his book of rules he can answer the questions with the appropriate Chinese symbols. He passes these answers back through the slot. To the people outside the room it seems that Searle can understand Chinese; after all, he's been given a story and has answered questions about it. The point of the exercise is that he can't; he's just following instructions.

Searle argues that a computer program is the same thing. To produce the right answer, it doesn't have to understand anything either. It merely has to follow a set of rules.

Thankfully for our purposes, reasoned philosophical arguments like this don't wash with most AI researchers, although it doesn't half make them

Brilliant mind

Although Alan Turing never used the phrase 'artificial intelligence', he laid the groundwork for much that would follow. It is only in more recent years that his contribution to computers and AI has been fully appreciated. His brilliance at logic and code breaking helped to save thousands of lives during the Second World War. The man himself has been variously described as shabby, eccentric, awkward and troubled. Perfect genius material.

While still at school Turing showed an understanding of Einstein's theories and despite the best efforts of the British public school system managed to land a place at King's College, Cambridge. His work on mathematical logic laid down the basic rules of computing and became the blueprint for the first digital computers. After a spell in the States he returned to Britain and started secret government work on codes and cyphers, the ultimate aim of which was to crack the German Enigma codes, used by the U-boats attacking Allied shipping in the north Atlantic.

Turing worked long hours and his eccentricities were legendary. Eyewitnesses often reported seeing him cycling in a gas mask, apparently to help his allergies, or running with an alarm clock strapped to his wrist (the reasons for this are unclear). He was also gay, a fact he never particularly tried to hide. That, at a time when homosexuality was illegal, demonstrated immense strength of character.

Turing's war efforts remained secret for decades but did stand him in good stead to contribute to the design of the first true electronic computers. In the years that followed, he came up with concepts that would later form the basis of research into neural nets (more on those later). He also contemplated the philosophy of machines that could learn, and gave lectures and radio broadcasts on how computers could change society. What made Turing even more remarkable was his ability to turn his mind to a broad range of scientific disciplines, such as chemistry and biology.

In March 1952 he had a brief affair with a Manchester man who later stole some money from him. Turing made the mistake of reporting this to the police, who duly arrested Turing for homosexual activities. He was sentenced to the humiliating treatment of oestrogen injections to 'cure' him of his homo-sexuality. In his later years he was shunned by the same authorities he had helped win the war and lost his security clearance. He died on the 7 June 1954 after eating an apple laced with cyanide. A verdict of suicide was recorded by the coroner.

Incidentally, you'll find numerous articles on the web suggesting that the Apple Computers logo was conceived as a tribute to Turing (it features an apple with a bite taken out of it). Having asked the company and received a somewhat ambiguous reply, we're not convinced this is the case. Shame really.

angry. They contest that although the person inside the room doesn't under-stand Chinese, the 'system' does – that's to say, the room, the rules, the cards *and* the person.

BUILDING A BRAIN

Perhaps the most logical and foolproof way of designing an artificial intelli-gence would be to copy the way the brain works. There's one tiny flaw in this approach: scientists don't quite *know* how the brain works. They do know what it's made from – nerve cells, known as neurons. The beauty of the human brain is that although the individual components might be relatively simple, the brain itself is certainly not.

Neurons have all the same basic internal structures as any other animal cell, including a nucleus (holding the genetic material, DNA) and a mito-chondrion (the powerhouse). All these are contained in the main body of the cell. A series of small branches called dendrites lead out from the cell body, as well as one longer branch called the axon – this is the actual nerve 'fibre'.

To our minds a neuron looks like a stick man with spiky hair. The man's head is the nucleus, his hair is the dendrites, and his body is the axon (well, it works for us). Individual fibres are bundled together to form nerves and can be incredibly long, at least a metre in the case of the fibres that run from the base of our spine to the tips of our toes.

When a nerve is stimulated, an electrical charge is passed along the axon. At the end of the axon the cell branches, each branch culminating at a junction with an adjacent cell. Unlike the connection in a couple of wires, the axon isn't physically joined to other neurons. Instead, when the electrical charge reaches these junctions, or synapses, chemicals are released which pass across the tiny gap between the cells.

These chemicals, called neurotransmitters, dock into receptor sites on the dendrites of adjacent neurons and another small electrical charge is generated. Each nerve cell has many hundreds of dendrites and will only fire off its own electrical impulse if a certain number of them are stimulated. One dendrite alone isn't enough.

So what about the brain? It has around 100 billion neurons (that's 100,000,000,000). Each one of those neurons has hundreds of connections with neighbouring cells. As we go through life, the resulting network of neurons is constantly changing – new cells are joining, old cells are dying. As you read this, the inside of your brain is being rewired. Makes you think, doesn't it?

The closest thing computer scientists have to neurons is logic gates, the basic building blocks of a digital computer. Constructing Nigella's artificial brain is likely to involve quite a lot of silicon (and you'll probably need some silicone if you want your robot to be anatomically correct). So is it as simple as 'How many silicon chips do we need to achieve intelligence?' Here AI researchers are divided. Do you cram loads of chips into Nigella – or Jeeves – and hope for the best, or do you start with a simple Nigella and tutor it? The two approaches are known as 'top down' and 'bottom up'. A sentence surely crying out for crude innuendo.

TOP DOWN

Early efforts at artificial intelligence are best described by the phrase 'wildly optimistic'. It was a time when computers looked like computers ought to. If they didn't take up half an aircraft hanger and use the power of a small city then they weren't worth having. They consisted of racks of electronics and flashing lights.

Every so often there was a bleep and clatter as a line of ticker-tape spewed out of the side. The future, as seen in the fifties, was all stainless steel and nuclear-powered. We'd all be leading lives of leisure while machines buzzed around doing all the messy stuff. Think Robbie the robot in *The Forbidden Planet* making cupcakes and taking out the trash at the same time. The man of tomorrow would spend his days lounging in the garden smoking his pipe while his wife discussed knitting patterns with her friends (actually she was having an affair and spent most of the day drinking herself into oblivion, but you get the idea).

The reality is somewhat different. Few men smoke pipes, stainless steel is too expensive and computers are a good deal smaller than anyone ever imagined. Although people still have affairs. We now spend a large proportion of our lives sitting behind beige boxes grappling with the latest version of Windows. But in our 1950s dreamworld, electronic digital computers were only a few years old. Despite their colossal size, power consumption and cost, they were already proving their worth for businesses, the military and scientific calculation. Rather than just using computers to do the accounts or work out a missile trajectory it seemed an obvious next step to apply them to thinking.

The first person to use the phrase 'artificial intelligence' was John McCarthy in 1955, when he proposed a summer research project at Dartmouth College, an American Ivy League institution. It took place the following year, drawing together some of the pioneers of AI and laying the framework for future development.

American researchers Herbert Simon and Allen Newell unveiled what's now widely considered the first true AI program in 1956. Called Logic Theorist, it could solve mathematical problems using the rules of logic. Their next project, which they developed with another researcher, Cliff Shaw, was grandly named The General Problem Solver. Again, by using a set of rules, the program could solve logical problems – manipulating shapes or playing noughts and crosses. Unfortunately, it couldn't solve general problems like, 'Where are my underpants?', because it had no knowledge of the real world, let alone your smalls.

Highly illogical, Captain

At its most basic level, a computer is a relatively simple series of switches called logic gates. Like neurons, these logic gates have inputs and outputs. Humans generally count in decimal, units of ten, probably because we have ten fingers. Digital computers work in binary and count in units of two. One to ten in binary is 1, 10, 11, 100, 101, 110, 111, 1000, 1001, 1010. This means that the inputs and outputs can only be '0' or '1' and nothing else.

These binary 0s and 1s are referred to as 'bits'. Traditionally there are eight bits in a byte, and 1,024 bytes in a kilobyte. A megabyte is 1,000 kilobytes, or 1,024,000 bytes, which means a one-megabyte computer can cope with eight million, one hundred and ninety-two thousand 0s and 1s.

Which is quite a lot.

But don't worry about the numbers, it's what the gates do to individual 'bits' that is the important thing. Each different type of gate uses something called 'Boolean Logic', named after the man who invented it in the 1800s, George Boole. *Star Trek* fans might find it easier to think of it as Vulcan logic, because it's exactly the same thing.

There are three basic types of logic gate. The first is called a NOT gate – stick something in and the opposite comes out. Give it a 1 and it'll give you a 0 and vice versa. The next is called an AND gate. It has two inputs, both of which have to be 1 for it to produce 1 as an output. Other combinations of 0 and 1 only produce 0. Finally, there's the OR gate. With an OR gate if *either* of the two inputs are 1 it will output a 1.

There are several other types of gate, which are all variations on the three basic ones: NANDs, which combine NOTs and ANDs; and NORs, which combine NOTs and ORs. There are also XORs and XNORs. But you won't be tested on this later, so don't worry too much about the detail.

The point is, depending on the type of logic gate, different inputs will produce different outputs. Listing what each particular one does to each combination of inputs would be somewhat tedious. Particularly as a gate on its own is not very useful; it is only when you stick a load together that the network starts to become interesting.

For instance, a simple addition like 0 + 1 can be done using an XOR gate. With this type of gate, feeding in 0 to one of the inputs and 1 to the other produces an output of 1. Which is fine until you want to add 1+1. In decimal 1+1 = 2, in binary 1+1 =10. Because the XOR gate can only produce a one-bit output, it gives the answer as 0. Rather than lose the 1 altogether it can be carried over to another gate. It is the same as adding numbers up in decimal. On reaching ten, the 0 goes in the first column and the 1 is carried over to the next.

To build a basic adding machine using logic gates involves combining several different gates together. In a computer the 1s and 0s are actually pulses of electricity. Gates either let the electricity through to produce a 1 or block it to produce a 0. The logic gates themselves are microscopic transistors etched into silicon. Each chip contains millions of them, all processing 0s and 1s. Together these can represent the letters on a page, the numbers on a spreadsheet or the pictures on a website. Memories can be created in them – and just as easily lost for ever. All each computer component is doing is manipulating 0s and 1s. But as a network of neurons produces a brain, a network of logic gates would seem to be a good place to start to build an *artificial* brain for a domestic robot.

A WORLD OF ITS OWN

The fundamental problem with top-down AI is that you have to program all the knowledge in before you can get a sensible answer out. The system can't 'learn' common sense; it has to be told. To work in any meaningful way, top-down AI needs knowledge and rules in order to apply itself to a given situation. Understandably, many researchers baulk at the idea of programming the world and every possible situation encountered into a machine. Instead, they create virtual worlds, or 'domains', in which their artificial intelligences can frolic at pleasure, if indeed they have an idea of pleasure. Which they don't.

One of the first successful examples of this sort of AI was a program called

SHRDLU (see the box on page 51 for how it got its name). Devised in the late sixties by computer scientist Terry Winograd, it was designed to understand conversation and interact with humans, albeit to a limited extent. The world of SHRDLU consisted of a room full of blocks – squares, rectangles and pyramids. The user would type instructions in regular English sentences, this is termed 'natural' language, as opposed to computer language, and the program would move the blocks around. You could also ask it questions about its 'world' and it would spell out a reply. In a demonstration, the researcher asked it, 'What is the pyramid supported by?' and up on the computer screen flashed 'The box'. You can download a version of the program to this day, and while away many a happy afternoon in SHRDLU's virtual world. Or you could have a conversation with a real person. Hey, it's your life.

Although it is limited to its own peculiar world, SHRDLU demonstrates two of the main attributes any domestic goddess needs to possess (no, not those two): an awareness of the world around it and an ability to interpret instructions. If all you do is ask the program to move blocks around, it appears that the computer is actually reading what you type in and understanding it. On the face of it, not much different from a human.

If we're ever to get our robotic domestic goddess off the drawing board, natural language processing would seem to be pretty important. In fact, some researchers believe that understanding natural language is the key to mastering artificial intelligence. We need to be able to give it instructions, 'Tidy the kitchen', say, and know our Nigella (Mark 5) will tidy the kitchen. We don't want it slumping on the sofa watching MTV. That's what we'll be doing.

In order to work out what someone is saying, a computer has to know what the words in the sentence are, how the order of the words gives the sentence its structure and the context of those words. Within this it has to understand the numerous ambiguities in language. 'Duck, it's a ball' could either be an instruction to take cover or a conversation with a wildfowl about a posh party. A computer has to know both meanings of the word 'duck'. It also has to know that ducks can't understand conversation (unless it's reading

SHRDLU

The creator of SHRDLU, Terry Winograd, admits that when he tried to come up with acronyms for his program he didn't think any were very good, so he used SHRDLU instead. It looks like an acronym but isn't.

Like many things in the world of computers, it belongs to a bygone era. Until relatively recently (the 1980s) the words in newspapers were put together for printing using mechanical linotype machines. Workers used a special keyboard to produce brass moulds, or dies, of each letter. When a line of dies was complete it moved to a second part of the machine, where it was cast in molten metal. When cooled these linotype slugs were used to make up the page for the printing machines.

On a linotype machine the first column of letters read 'ETAOIN', the second 'SHRDLU'. The order of letters was based on how often they were used, 'e' being the most frequently used, 't' the next, etc. Because this was a mechanical process there was no delete key. Instead, if the operators made a mistake they would fill lines with things like 'SHRDLUSHRDLUSHRDLU' so the proofreaders could pick up the error. Unfortunately, these lines of gibberish would sometimes end up in print. Something of a conversation starter if you're ever stuck in a lift with a computer scientist.

Incidentally, while we're on the subject of random letters, the QWERTY keyboard dates back to the earliest typewriters. With a mechanical typewriter, hitting a key causes a spring-loaded lever with the letter cast at the end to be catapulted towards the ink ribbon. The paper is positioned behind the ribbon so an impression of the appropriate letter is left on the page. Because the levers corresponding to each letter are arranged in a tightly packed semi-circle, when adjacent keys are hit one after the other they have a tendency to jam. The QWERTY keyboard is designed so you are unlikely to want to type adjacent keys in rapid succession. In English W never follows Q for example. With computers, of course, this is completely irrelevant, but all attempts to change the layout of the keyboard have faltered. If you hate QWERTY, blame the Victorians.

a children's book) and be aware that balls can fly through the air and might pose a danger.

Take every possible sequence of words in every possible intelligible sentence and you can see the problem. Any natural language-processing software has to be able to break a sentence down into smaller and smaller pieces, identify the various components (verbs, nouns, etc.) and from that work out what is being said. To function realistically the computer has to have a knowledge of the human environment and what is and isn't possible.

There are dozens of research projects going on in the world into developing these sorts of human/computer interfaces – with varying degrees of success. They range from translation programs that overcome language barriers for emergency workers to the incredibly annoying paperclip help function on Microsoft Word. Natural language processing could well be the key to passing the Turing test. Of course, just because a computer seems to have understood you, doesn't mean it has.

EXPERT KNOWLEDGE

For a top-down artificial intelligence to function properly it needs some knowledge of its environment. SHRDLU, for instance, was an expert on its little block world. If you wanted a pyramid moving on top of a square, SHRDLU was your man. Other so-called 'expert systems' function in a similar way and are proving their worth on a daily basis. The very definition of an 'expert' implies that not everyone has that particular knowledge, so there are only so many experts to go around. So while SHRDLU's world, or domain, consisted of blocks, the domain of another expert system could be, for example, the diagnosis of cancer.

The programming of this type of software is based on human knowledge and not just that of one human. In theory you can program in everything that is known about a particular disease based on scientific research, articles in journals and interviews with human experts. It's the latter which is arguably the most important, because you are giving the system human experience.

Coupled with natural language processing, expert systems can be a much more useful tool for doctors than simply consulting a textbook.

An example is a program called FocalPoint, which is expert at looking for signs of cancer. It is being used in the United States to examine cervical smears microscopically for anomalies. Unlike a human operator, it arrives fully trained and does not get tired or lose concentration. Expert systems have also been successfully applied to the techniques of behavioural therapy. Several psychiatric hospitals use them to help treat phobias and drug and alcohol addiction. The computers interview the patients and lead them through their treatment programmes. In clinical trails they've proved to be more than a match for their human counterparts.

THE CYCLISTS

The bigger the domain an expert system operates in, the more information it needs to function properly. To work in the real world, or at least your living room, Nigella or Jeeves needs a knowledge of the world. To be useful in any domestic situation it needs to know about people, rooms, furniture and where the beer is kept. In short, Nigella needs to be programmed with common sense. So is it possible? One group of scientists certainly thinks so.

Since 1984 computer researchers in Austin, Texas, have been attempting to put *everything* into a computer. In creating Cyc (from 'encyclopaedia') they've broken the world down into categories and worked out the relationships between those categories. Cyc knows, for example, what a liquid is, why people need to drink and where you can get a beer. Cyc is centred on a database of at least one million logical 'assertions', each entered manually by researchers. The rules describe the world around us in the way that perhaps an alien would, looking down on Earth from an orbiting spacecraft.

Cyc is not just programmed with hard facts. Its database includes such aspects of life as human emotions and motivations. It can't explain *why* people fall in love but might be used by a dating agency to cross-reference questionnaires and match up suitable couples. The program is already being used to spot errors in company databases and make web search engines more accurate.

But once you have bought the latest version from the Cycorp labs, you can't teach it anything new. Cyc doesn't learn; it only applies what it already knows. Because it operates within statements of logic, it won't necessarily be able to find where you left the car keys or understand the offside rule.

This is one of the fundamental problems with top-down AI: you have to put a huge amount in to get anything vaguely intelligent out. Cyc has taken almost 20 years to develop and is by no means perfect. As philosophers will doubtless point out at great length, we are more than just a sum of our knowledge. Having an understanding of logic does not explain all, or, indeed, probably most, aspects of human behaviour. Perhaps a true AI should be like a human, with feelings, emotions and a liking for Marmite. Or as this 'intelligent' spellcheck would have us write, marmot.

BOTTOMS UP

Unlike a top-down AI, humans don't come pre-programmed to cope with life. For at least the first few years of existence we are totally dependent on others. We need to be provided with food, shelter and clean nappies. But from the moment we are born we do start learning about the world. It is this ability to learn that comes pre-programmed, not the information itself.

Professor Alan Turing talked of developing a computer child that would 'learn' in the same way as a human. To realise this dream, many AI researchers are looking to biology for inspiration. When we learn, the connections between our neurons change, so if we want to build a brain, developing some sort of *artificial* neuron would seem to be a good starting point. In 1958 American Frank Rosenblatt came up with the first system that could be trained. Called the Perceptron, its artificial neurons functioned in more or less the same way as a nerve cell.

This is how it works. Each unit of the Perceptron receives several inputs. When these are added together if they overcome a pre-set threshold they produce an output (in the same way that a nerve cell only fires when several dendrites are stimulated). By changing the relative importance that each unit assigns to each particular input, the Perceptron can be *trained* to produce the right answer. Researchers successfully taught networks of artificial neurons to recognise patterns in particular letters and numbers. Give it a 'P' and it'll tell you it's a 'P'. Trust us, this is useful.

CAN I HAVE A P PLEASE, BOB?

Imagine the letter 'P' drawn in black on a large sheet of white paper. Now point a video camera at it. The camera will break that 'P' down into tiny blocks or pixels. Each pixel represents a proportion of that letter. Some will be black (the letter), some white (the space around the letter). The camera converts those pixels into electrical signals so that the strength of the signals depends on whether the pixel is black or white. For this example, the white pixels produce a low current, the black a high current. These currents become the inputs to the artificial neurons of the Perceptron.

Each element in the Perceptron is fed all the inputs (the currents from all the pixels) and is pre-programmed with rules giving each one a relative importance. Input 1, perhaps the white pixel in the bottom right-hand corner of the square, might be assigned a low importance. Input 10, let's say the top of the vertical line, might be given a high importance. Each artificial neuron adds the inputs together, and if they overcome a certain threshold then that particular unit produces an output. Taken together those outputs can be converted back into a TV signal as the Perceptron's interpretation of the letter 'P'.

What's clever about this system is that you don't have to get the starting rules right. Say you feed in the letter 'P' and the Perceptron comes up with 'R'. Because this is wrong, the error is fed back into the network and the relative importance of the inputs are adjusted slightly. The process continues until the Perceptron gets it right. This feedback mechanism is automatic and the 'training' for each letter takes fractions of a second. A basic Perceptron network can be taught to recognise letters in different styles, written at different angles, by different people.

Researchers soon worked out that you didn't even need to build artificial neurons; you could emulate them in software and write learning programs on ordinary computers. By the late 1960s it looked as if computers that could 'learn' would revolutionise technology. So why are we all using conventional PCs when we could all be using Perceptrometers, which could maybe glow in a throbbing, futuristic way? Sadly, the Perceptron couldn't perform the sorts of basic logic problems that an ordinary binary computer, and the human brain, does so well. But in our quest for Nigella, all is not lost.

CASTING THE NEURAL NET

A Perceptron is made up of artificial neurons and can be taught to recognise shapes. This type of learning machine is known as a single layer network. Each artificial neuron is only connected to inputs and an output. Scientists realised that by adding extra layers of artificial neurons they could not only overcome the logical shortcomings of the Perceptron but they could create a much more powerful learning machine.

At its most basic, a neural net is a load of Perceptrons stuck together. The net starts with an input layer, the output from that layer becomes the input to the next, the output from that becomes the input to the next and so on, until finally there's an output layer. The layers in between the input and output are known as hidden layers. There's a good reason for this. Scientists have no idea what's going on in them.

Apart from the extra layers, a neural net differs from a Perceptron in the way that errors are corrected. During training, if it doesn't come up with the right answer, the error is fed back through the whole network and the relative importance assigned to each input is adjusted accordingly. This is called back propagation and is a way of adjusting the inputs throughout each one of the hidden layers.

Neural nets are extremely good at recognising patterns. Connect a camera up to them and they become artificial vision. You can buy them on a chip or run them on your computer. They can be used to recognise handwriting and car number plates, and even to pick out faces in a crowd. You can see why governments fund their development.

Artificial neural networks do have their limitations. The human brain has around 100 billion neurons, each one connected to hundreds of others. Even the most complex neural net is nowhere near as complicated. More fundamentally, although artificial neurons come close to the real thing, they still work in a different way from the human brain, as we don't use back propagation to learn. And whereas a neural net is *designed*, we are very much a product of evolution.

HOW TO BREED A DOMESTIC GODDESS

If someone set out to design a perfect being, they probably wouldn't come up with a human. We're inherently unstable for one thing – not because we cry at the end of dubious Julia Roberts films or threaten the world with nuclear annihilation (obviously not everyone does that). We can barely stand upright. Our brain has to make constant adjustments to our posture so we can keep our balance. Four legs would be much more stable, but then we wouldn't be able to use our hands. Perhaps an ideal human would have four legs, two arms and a pair of wheels. But remember we weren't *designed*, we evolved.

So rather than trying to design artificial life, some researchers are learning from biology and trying to evolve it. And while it has taken billions of years to get from a blob of slime to a human (depending on the human) computer scientists are developing software that can evolve in seconds. Being scientists, they give these types of program a scientific-sounding name: genetic algorithms. What they're doing is handing computers a set of evolutionary rules. While researchers almost certainly won't be able to evolve a Mark 5 Nigella from a pile of electronic components, understanding how evolution works might help Nigella to think. If we humans were able to evolve the ability, there doesn't seem to be any fundamental reason why an artificial life form couldn't as well.

The driving forces behind evolution in the biological world are natural selection – often referred to as survival of the fittest – and, of course, sex. Long before an organism, which we'll call Dave, can even contemplate exchanging sophisticated witticisms over a cup of Gold Blend, Dave has to survive long enough to make that evening possible. Having survived, Dave's pre-programmed goal in life is to have lots of sex and pass on his genes to a new generation.

For most plants and animals, both male and female, life really is that simple. Those that get to reproduce have the ability not just to survive; they're also likely to have attributes that give them the edge over their competitors. In a flower this could be pollen that's more attractive to an insect; in a bird it could be a particularly bright plumage or a loud mating call that makes them more desirable. Some animals, of course, hedge their bets by mating with loads of partners (see Warren Beattie), others have and hold for a lifetime (see the Queen) and there are some that try to have it both ways (see Warren Beattie again). Biologists have described certain types of penguin, for example, as socially monogamous but sexually promiscuous. Whichever strategy a particular species goes for, the goal is the same: passing on genes. With half the genes from one parent and half from another, at least some of the desirable traits will be inherited by their offspring.

Another interesting concept in evolution is genetic variation. Although all humans are the same species, we come in an enormous range of colours, shapes and sizes. Some of this variation is determined by our genes, some by the environment we live in. The same applies to every other plant and animal. Over evolutionary time the desirable characteristics from this pool of genetic variation are more likely to be passed on to future generations.

There's also a random element at work – the wife-swapping party of the evolutionary process. Every so often, a gene can change its function or a new gene can start to operate. If these random mutations are desirable they will most likely be passed on to the next generation.

Genes are made up of DNA (see page 10), which is a string of biological code. Instead of DNA, a genetic algorithm uses computer code (zeros and ones) to achieve much the same thing in a virtual environment. Rather than individual organisms competing to reproduce, researchers unleash a variety of different programs and the best ones survive. To 'live' the programs need to be able to accomplish specific tasks, or at least get the closest to their goal. Those that do, get to cross-breed with other successful programs in a kind of virtual sex. By repeating the process over hundreds of generations and introducing the occasional mutation, the best software should evolve.

Like the neural net, a genetic algorithm is another way a computer can 'learn' to do something. Assuming your virtual gene pool has got some reasonably sensible code to start with, in a few generations (a couple of hours say) it should have reached its goal. But unlike a neural net, a genetic algorithm doesn't need to be carefully taught. It can evolve unsupervised.

At the University of Sussex, scientists have gone one step further. Rather than just limiting themselves to software programs, they're developing what's called evolvable hardware – in this case, computer chips that can be rewired according to evolutionary rules. It's a relatively new area of research because the technology is only just becoming available, but it raises the possibility of designing full-size robots that change as they mature, just as we do.

Which brings us to slime.

SLIME

As life forms go, slime is about as primitive as it gets. Slime mould isn't a plant, an animal or even a fungus but inhabits its own completely separate kingdom. Neither is it much to look at – a white or yellow film found on wet grass, not unlike congealed scrambled eggs. Gardeners like slime mould because it keeps the soil in good shape, which is interesting but not that relevant. AI researchers like it because it appears to be quite intelligent.

In its most basic form, slime mould is a single-celled amoeba-like organism. But under certain conditions individual cells come together to form a much larger living entity, and when this happens it can do some remarkable things. In a paper published in the science journal *Nature* in September 2000, some

Japanese scientists, led by Toshiyuki Nakagaki, discovered that slime mould could find its way through a maze. If you've ever been to Hampton Court, you'll know this is no mean achievement (and a somewhat sobering thought to realise that slime might be better at it than you).

The researchers placed some slime mould on a 30cm-square maze with food at two of the four exits. Time and again the organism managed to discover the most efficient route between the two food sources, thereby solving the puzzle. Not only does this show that an organism doesn't need to be complicated to act in an 'intelligent' way but it is also an example of how simple individuals can work together to achieve relatively complex behaviour.

Take ants. Each individual ant is quite a basic organism, and although far higher up the evolutionary scale than slime mould, not a great deal brighter. Although they've got a pretty impressive range of sense and movement for something so small, when they work alone they're not very effective. But together a colony of ants can take on the world. They can move objects hundreds of times their body size, co-ordinate food supplies, remove waste, design and build impressive nests and annoy the hell out of you when they get at the sugar.

When biologists first started looking at creatures like ants that display this swarm-like behaviour, they likened them to human societies. They figured that the queen must be in control, co-ordinating the workings of the nest and all its inhabitants. But comparing animal behaviour with human behaviour is fraught with pitfalls.

If you think about it, this top-down approach couldn't really work. How could one individual control thousands, even millions, of others? The queen would have to have either a massive brain or a vast bureaucracy to keep everything under control (and presumably some spin-doctors to keep every ant 'on message').

Instead, each ant obeys simple rules governing not only its basic functions but also its interactions with others. By communicating with each other using chemical trails called pheromones, the colony as a whole can pass messages such as 'There's food south of here' and exhibit complex behaviours. The study of this sort of thing is called emergence. The queen is merely a baby ant factory, vital to the colony but not in charge. In fact, no one is in charge. That's the beauty of the 'self-organising' system. Emergence is another case against top-down systems. Why try to build a robot with a giant brain when a few basic rules will do the trick? For our purposes, the great thing about emergence is that it works for robots too.

BIMBOTS AND HIMBOTS

Let's put aside our dream of a multi-purpose, multi-functional intelligent domestic goddess for a moment and consider a simpler option. If you think about it, to get around the house a robot doesn't need to know what a wall actually *is*, it just needs to know it can't go through it and needs to change direction. A simple rule that tells it to turn if it encounters an obstacle might do the trick, rather than a complicated set of instructions involving bricks, plaster and paint. So maybe instead of a domestic goddess we should be aiming for a less intelligent version of Nigella, a domestic bimbo – or bimbot if you will. Or, if it's Jeeves, a himbo or himbot. Or maybe a swarm of bimbots and himbots?

Scientists at the University of Reading have tried just this. They set up a series of experiments using seven simple three-wheeled robots. Named the 'seven dwarfs', each one was about the size of a boot. They were contained within a paddock on the laboratory floor and were technically pretty simple: a few basic sensors and a couple of motors all connected to a computer processor. They were also equipped with the ability to communicate with each other using infrared transmitters and receivers. Like ants, they were programmed with a set of simple rules and a goal – such as 'Travel in the same direction'.

When first switched on they wheeled around randomly using sonar to avoid obstacles. But in order to achieve their pre-programmed goal they ended up 'speaking' to each other using a series of infrared flashes. What's so clever is that they weren't told to do this. The behaviour emerged. Invariably the dwarfs managed to do what they were asked without being told how. At times, for example, a leader would emerge and the rest would follow. The leader could be any one of the robots because they were all created equally, with the same

Management speak

Emergence is part of an area of research known as complexity science. Not only because it's bloody difficult. In broad terms complexity science describes what happens when a load of individuals, each with their own rules, start interacting together. In nature, ants are not particularly complicated but an ant colony is. The same science can be brought to bear on human society when you're dealing with the interactions between people. The idea is to try to make some sense of the rules that the components of the system operate under. That in turn can help you understand what's going on in the system as a whole.

Complexity science can be used to control traffic in a busy city. In theory the system consists of lots of individuals in charge of machines following a set of rules (the Highway Code) as they move around. Of course, if it were that simple there would never be any traffic jams. In reality the system consists of lots of individuals with tempers, mobile phones and babies on board. These individuals may have only a vague idea of geography and have forgotten or chosen to ignore the rules of the road. This is when complexity science gets really interesting. If you can start taking those sorts of factors into account you can build a simulation of a traffic system that actually works. Of course, when you try it for real it rarely does, but then that's the challenge. And complexity science is not only useful for working out the timing for traffic lights; it can also be applied to entire organisations.

In recent years some business gurus have got very excited about the potential of complexity science. As we've already mentioned, a top-down approach can be incredibly cumbersome. Either the boss has to keep on top of everything that's going on or else he has to employ legions of middle managers to keep the company afloat. By adopting a more self-organising approach, a whole load of managers can be removed. After all, most of the workers in any organisation know exactly what they're doing and are quite capable of making their own everyday decisions. Unfortunately, once the management consultants have got their hands on it, self-organisation tends to end up as 'empowerment' and the managers simply get new titles like 'empowerment facilitator' – or they set up on their own as consultants.

Taken to its logical conclusion you could run whole countries using this way of thinking. One of the most interesting things about complexity science and emergence is that when individuals get together you can't always predict what's going to happen. Think of it as the scientific equivalent of the Eurovision Song Contest.

programming. At the end of the experiment, to paraphrase George Orwell, some robots were more equal than others.

To achieve their goal, the robots did not need to develop any sort of description of the world around them. Whether the seven dwarfs were displaying true 'intelligence' doesn't really matter. For decades AI researchers have been trying to build some sort of reasoning system into their designs. In experiments like the one at the University of Reading there is no attempt to do this; a bimbot simply reacts to the world around it and 'intelligent' behaviour emerges as a result. A robot doesn't have to understand why it's doing something or even exactly what it's doing.

Go online and you can buy a cleaning robot. There are quite a few out there – at prices to suit every budget (actually that's not strictly true, your budget has to be at least a couple of hundred pounds to contemplate the purchase of one). They don't have arms, legs, eyes – or in fact anything much. They simply scurry across the floor vacuuming up dust. But they don't know it's dust, they just vacuum. If they encounter an obstacle they turn round. In most cases they just bumble around randomly, eventually cleaning the entire room, until their battery runs out. With a few more sensors built in they can get an idea of where they've been and can move on to another part of the room. You can go out for the day and leave your little bimbot scurrying around the living room terrifying the cat.

What if you had several bimbots all working together or all working independently but interacting using simple rules? Like the seven dwarfs but with vacuuming nozzles. Rodney Brooks, director of the artificial intelligence

laboratory at the Massachusetts Institute of Technology (MIT) and something of an AI visionary, talks of a household 'ecology' of little robots, all cleaning and working together. They wouldn't all need to be the same size either; some could be insect-like, picking up fluff from around the bed and feeding it to larger robots that would be handling the bulk of the work. There doesn't even have to be any fluff there; the robots will carry on picking up anyway. They don't have to see what they're doing; they just have to do it. This is all perfectly possible, albeit a little disappointing. But if the idea of sharing your house with a dozen mini metal monsters isn't your vision of domestic bliss, let's return to our original premise – the lovely Mark 5 Nigella – or, if you prefer, our male model, Jeeves.

NIGELLA

Despite more than 50 years of research by some very bright, visionary people, there's no Nigella, no Terminator and certainly nothing like Data from *Star Trek* (who is even 'fully functional', although if you thought we were going to get into the sex lives of fictional TV characters you've bought the wrong book. That's the sequel). So how close are we?

What most science-fiction robots have in common is a humanoid appearance. Being able to interact with a robot through some sort of interface is important – which is where natural language processing comes in – but some argue that a *real* face is just as vital.

At a laboratory in Dallas, Texas, an AI researcher and former Disney employee, David Hanson, has built the face of the future. Called K-bot, the android head is made of a plastic polymer and has a repertoire of 28 facial expressions. With cameras behind its eyeballs and 24 tiny artificial muscles, it can follow a person with its eyes and mimic their facial expressions. K-bot can smile, frown, sneer and squint. The concept, known as social robotics, is designed to improve the interactions between humans and computers.

At the moment K-bot can merely copy human expressions. Future generations of android could learn to interpret them. As the humanoid appearance of the robot causes us to behave towards it as if it were another human, the robot

The evil within

In science fiction the robots are generally not the good guys. From the killing machine of *Terminator* to the, er, killing machines of *Westworld*, a theme emerges of intelligent automatons created by man that ultimately either bite the hand that feeds them or, more accurately, blow their creators' brains out. The artificially intelligent don't even have to be in humanoid form. The disembodied HAL-9000 computer in Stanley Kubrick's movie *2001: A Space Odyssey* is every bit as terrifying as the chillingly murderous android played by Yul Brynner in Michael Crichton's *Westworld*.

When Isaac Asimov created his laws of robotics, he considered the dangers inherent in building intelligent machines. His laws are designed to protect humans from their creations but so far they have proved to be largely irrelevant. Even if a scientist set out to build a gun-toting psychotic, they would find the technology simply does not exist. The worst an out-of-control bimbot can do is give you a sharp knock on the shin.

Researchers don't need to take the laws into account because the robots they create are barely powerful enough to answer back, let alone harm anyone. But now some scientists are beginning to take the machine threat a bit more seriously. Professor Kevin Warwick of Reading University thinks robots will ultimately decide that they are much better at things than humans and take over. Although Warwick is sometimes ridiculed for his beliefs, perhaps he does have a point. After all, if an android has the means to harm humans, through being equipped with a laser say, it doesn't need intelligence to use it.

A related view is taken by Professor Hugo de Garis, head of Utah State University's Artificial Brain Project. As AI machines, or 'Artilects' as he calls them, start to develop true intelligence, he suggests humanity will become increasingly divided. Some of us will back AI development, seeing it as the next step in human evolution. De Garis calls these people the 'Cosmists'. Others, whom he terms the 'Terrans', will be bitterly opposed. The resulting war between Cosmists and Terrans will kill millions and could culminate in the destruction of humanity.

So how close are we to android apocalypse? The United States military now regularly uses pilot-less spy planes. They are small, difficult to detect and can send pictures back from areas far too dangerous for conventional aircraft. If they are shot down it costs the US taxpayer a few hundred thousand dollars but not a life. At the moment most of these planes are flown by remote control. Using consoles in the Pentagon, the pilots can spy on remote parts of enemy territory. However, a new generation of spy planes is currently under development that will be completely autonomous. Once airborne, these will fly to a set of pre-programmed goals. Critics believe that a robotic spy plane is only a small step away from a robotic plane armed with guns and missiles. It won't be long, they argue, before people are killed by machines.

can copy our behaviour. It can learn what being a human is all about and we can teach it. Then our robot can clean the toilet, fill in the tax return and have a row about the mess in the kitchen. Which is what being a human is *really* all about.

K-bot certainly looks right. It is, after all, modelled on a real person, in the same way that a wax dummy is at Madame Tussaud's. At the moment it's not much brighter either. But other scientists have shown that only certain features need to be built into a robot for natural human interaction. Researchers at MIT have built a robot head called Kismet, which looks just, well, cute. With only a metal framework for a head, no skin, a pair of rubber lips and small cones for ears, the things that make it cute are the eyes. They're big and blue, with little eyebrows suspended a few centimetres above. By designing Kismet this way its creator, Cynthia Breazeal, has hit on just how important the eyes are to communication.

Although Kismet can speak, it doesn't utter anything intelligible. Although it can see, it doesn't know what it's seeing. Although it can hear, it doesn't know what it's hearing. Yet, remarkably, people can sit down and have a half-hour conversation with Kismet and interact with the robot in a human-like way. To achieve this sociability, Kismet is programmed with sets of rules and 'emotional' drives. For example, Kismet will pay attention to moving objects, turning to follow them. That drive is overridden by the appearance of skin – so if it sees a person it will turn towards them. Its drives include a boredom threshold, and in conversation it will follow a person's eyes, a gyroscope keeping its head balanced so that it moves in the same way as a human.

Kismet's speech may not make any sense but it speaks in the right places, holds eye contact when being spoken to and expresses emotion based on the voice inflections of the person talking to it. Kismet even has its own 'personal space'. Although this is based on the limitations of its cameras, it nevertheless functions in a human-like way. If someone is too far away for Kismet to see them properly, the robot will beckon them closer. If the person comes too close it won't be able to see their entire face and will physically draw back. So it acts in the same way as a human when cornered at a party.

The scientists working on Kismet compare its robot/human interactions with those between a parent and a child during the first year of life. There's no great intellect behind Kismet; it's not controlled by some vast machine, but it appears a good deal more intelligent than many of the top-down systems. The intentions behind the project are also very similar to those first described by Alan Turing more than 50 years ago in his article in *Mind*. The researchers at MIT are bringing up a baby.

COMING OF AGE

There are clearly problems inherent in the top-down approach to artificial intelligence. Unable to learn new things, a top-down Nigella, or Jeeves, would need to come pre-programmed to cope with every possible domestic situation. With a cold hard intellect, it would probably be an expert on the perfect house. But move the furniture around a bit and it would get horribly confused. Likewise, the simple cleaning robots we've talked about have their own problems. Accidentally leave the front door open and they'll start trundling off down the street dusting beneath the road signs.

A robot that can learn would appear to be a good thing to have, but you don't want to spend a year teaching it how to clean out the fridge, when a) life's too short, and b) you've never cleaned the fridge yourself. But what if you could buy a robot that had already been programmed with some real-world knowledge and could be applied to a new environment, your home, in just a few days? Something that utilised the best of the top-down systems (natural language processing, real-world knowledge, expert systems) and the bottom-up systems (neural nets, genetic algorithms, emergence, social behaviour)? If you could, you might eventually up with a genuine all-purpose Nigella.

There are a number of problems with this approach, not least that the respective proponents of top down and bottom up are as divided as ever. But some sort of system that learns based on prior knowledge would seem to make sense. You don't need a vast computational ability to wash the dishes, but you do need to be able to find the sink.

GETTING PRACTICAL

We don't want to discourage you from building a multi-purpose Nigella entirely, but there are a few other practicalities we should mention. To make any impact on the household chores our robot needs to be able to walk, manipulate objects and, of course, open a bottle of beer. Even the action of opening a beer is more challenging than you might think.

A fridge has no latch so opening it only requires an arm with a magnet on the end. Next, the robot needs some sort of beer recognition software (a neural net). Remember that you'd have to teach it to recognise every brand. Once it's identified the beer, it needs the ability to clasp the bottle without smashing it. As we'll discuss in the next chapter, the mechanics of grabbing the beer and opening it are not as simple as you'd think but could probably come pre-programmed. The beerbot then has to close the fridge while still holding the beer, requiring another arm. Finally, it has to trundle off and locate you, so it needs some wheels and motors. Alternatively, it could just squirt beer out of its body on demand. Not as refined but ultimately the same result.

The progress in artificial intelligence closely follows our understanding of how the human brain works. In fact the two areas of research feed off each

The wrong trousers

Most of us would probably want a robotic domestic goddess to look something like we do, vaguely humanoid in form. To operate in the home, Nigella needs to be a similar size and shape and have a comparable range of movements to us. Too tall and it'll keep smashing the door frames, too short and it'll never be able to reach the beer. And if our Nigella's going to have to navigate stairs or steps it'll need to be able to walk.

In the Aardman Animations production, *Wallace and Gromit*, our hapless heroes do battle with a pair of 'techno-trousers'. Designed to take the effort out of walking, the trousers (with Wallace inside) end up being hijacked by a villainous penguin. But designing techno-trousers is no mean feat. It has taken scientists decades to develop a machine that can walk.

We've already mentioned the amount of effort humans need to put into standing up as opposed to falling over. Likewise, there's nothing simple about moving around. The motion of walking doesn't just involve movement in our legs and feet. We also have to move our body, shifting our weight around so we don't topple. We need vision to see where we're going and a sense of balance to stay upright. It's also important to get information from sensors in our limbs so that we put just the right amount of pressure on our feet to avoid breaking our legs each time we take a step.

Most of the pioneering research on walking robots has been carried out by scientists at Waseda University in Tokyo. The first, named Wabot-1, took a few tentative steps back in 1973. It was followed by the piano-playing Wabot-2 in 1984, which could sight-read music. Both had legs and arms, but neither was particularly intelligent. Since then, researchers at what has now become the Humanoid Robotics Institute have developed sophisticated computer programs that make mechanical walking appear almost natural.

What started as relatively low-budget research has been taken up by the Japanese multinationals, among them Sony and Honda. The latter in particular has put enormous resources into building robots that can appear almost human in behaviour. The latest model is Asimo, which stands for Advanced Step in Innovative Mobility (presumably also a nod to Isaac Asimov). Asimo looks like a short astronaut in a space suit. At 120 centimetres high (about the size of a ten-year-old child), he can sit, walk, turn doorknobs and operate light switches. Like a human, the robot shifts his body weight as he moves, making walking appear smooth and natural. When he goes around a corner, his body tilts, moving his centre of gravity and maintaining his stability. Although Asimo has the right number of limbs, he does not have any toes. The engineers developing him discovered that the joints and balls of the feet were far more important to stability.

To all intents and purposes Asimo appears human. In fact, although the technology inside him is remarkably sophisticated, Asimo is remotely controlled. The robot relies on a human operator, leading Rodney Brooks at MIT to describe it as an 'expensive puppet'. Nevertheless, in the short term this research into walking will benefit amputees and those with certain disabilities. Combined with some artificial intelligence, this technology could make a true domestic goddess a reality.

other – through developing interactive robots, scientists can glean valuable information about the way our own minds function. At the moment, emulating human behaviour is extremely difficult, particularly when no one knows how humans think. This is why some people have contemplated a domestic goddess that combines the practicalities of a robot with the common sense of a human. By using a real human.

DOMESTIC SERVICE

While the mechanics of robotics are undoubtedly tricky, it is the brain itself that makes the difference between a bimbot and a Nigella, a himbot and a Jeeves. If only a human brain could be transplanted into a machine then all our problems would be solved. One way of doing this is to use a robot remotely controlled by a human operator. It's ironic that the concept, which doesn't require the robot itself to have any actual intelligence, has been described in detail by Professor John McCarthy, the man who first came up with the term

AI. He calls it 'teleservice'. In an essay on the subject he sees the robot and the operator being on opposite sides of the world.

In practical terms, the worker would operate a console in a computer centre in a developing country. He or she would be able to see your house, move around it and touch things within it. From your point of view, the robot would clean, cook, make the beds and terrify the cat (maybe you should think about getting rid of the cat). In the same way that many telephone call centres now operate out of low-wage countries, a robot operator would cost a lot less than a housekeeper would in the West.

The technology required already exists. Nuclear power plants use remote-control robots to handle radioactive materials. They can manipulate objects with perfect precision. So as not to crush things, there's a feedback mechanism, whereby the operator can 'feel' what the robot is doing (we'll talk more about this in Chapter 6). Obviously, any such a system would require an enormous amount of data to be transmitted back and forth. With broadband fibre optic links, this is now perfectly possible.

It wouldn't take long for your teleservice agent to become familiar with your house and what needs doing. With common sense behind it, the 'robot' would be able to work things out for itself. At the end of the day the worker goes home to their family and your robot folds itself up in the cupboard.

There are obviously some ethical issues to get your head around – like pay, conditions and sheer guilt. You would be letting a person with relatively few material possessions into your house, which in comparison to theirs would presumably be pretty opulent. That argument is perhaps somewhat naive. After all, most people in the developing world will have seen Western TV programmes such as the gritty soap *EastEnders* so will have an idea what to expect. The chances are your life is not nearly as interesting as the everyday murders and mayhem of Albert Square. If you had the same teleworker every day, you might well build up a rapport with the man or woman behind the metal.

McCarthy talks about extending the service to businesses such as hair-dressers – where your stylist is at a computer terminal in Jakarta. This might sound bizarre, but experiments with telesurgery are already showing promising results. Specialist surgeons can operate on patients thousands of kilometres away by remote control. And with a scalpel-wielding robot, there's a lot more at stake than just a dodgy perm.

TOGETHER IN ELECTRIC DREAMS

At the moment some basic fundamental problems have to be overcome before an all-purpose robotic domestic goddess becomes a reality. Current technology is certainly up to building a robot body, just not much intelligence to go with it. No brain, no gain, as they say. There are artificial intelligence laboratories all over the world in which scientists are working on all aspects of AI, but in Utah there's a researcher who's tackling the problem head on.

Professor Hugo de Garis calls himself 'the planet's pioneering brain builder'. He uses a combination of neural nets and genetic algorithms to mimic evolution. He's aiming for a machine containing about 1 billion artificial neurons. Although this is still only around one per cent of the capacity of the human brain, he's of the view that artificial brains will surpass the intelligence of humans in the not too distant future. De Garis bases his claims on Moore's Law. Named after its creator, Gordon Moore, the co-founder of Intel, this law states that the capacity (memory or processing power) of silicon chips doubles every year or so. As a result, de Garis says, 'Artificial intelligence is evolving a million times faster than human intelligence.'

As we've seen, artificial neurons still have some way to go, but de Garis is certainly optimistic. He's not alone in predicting that all-purpose domestic robots will make it on the scene within the next 30 years. Of course, many scientists were saying exactly the same thing 50 years ago. In the meantime, teleservicing might be your best bet. If the thought makes you uneasy, it's probably best to settle for a few relatively stupid but practical labour-saving devices. A fleet of bimbots, a beerbot and a dishwasher. We're afraid our version of Nigella is still some years away. As for the fully functional Mark 10 model, that remains a distant fantasy.

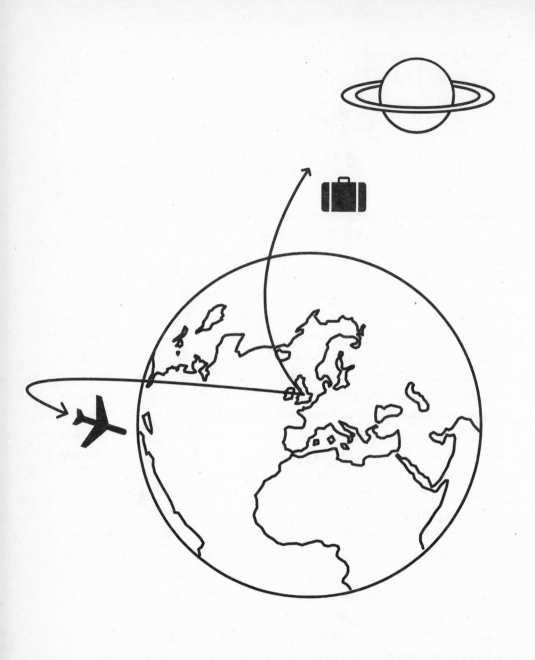

CHAPTER THREE

How to Avoid Commuting

There's no other way to put this. Commuting sucks. Whether it's by plane, train or automobile, commuting is one of life's tiresome necessities. If you doubt us, think again, because you don't need to travel to work on a daily basis to appreciate how any regular journey – be it to the shops, the in-laws' or for the school run – eventually becomes a chore.

Unfortunately, commuting becomes even more painful once you've purchased your very own Nigella or Jeeves, as the addition of a domestic robot to the household will have given you a taste of all the fun things to be done with your new-found spare time. Luckily, there is a potential solution ahead, even if it may, at first sight, appear to lie within the realms of science fiction. Don't worry if you can't tell a tricorder from a warp drive, you honestly don't need to be a *Star Trek* fan to appreciate the wonderful, time-saving concept of teleportation. But of course it helps.

The original series of *Star Trek* is set in the twenty-third century. It is a distinctive vision of the future in which interplanetary space travel is routine, injections never require a needle and the women all wear tight miniskirts. There are many memorable technological gadgets, and none more so than the device used for transportation – because when it came to commuting Captain James T. Kirk and his colleagues had away missions sorted. Crew members of the starship Enterprise would simply step into the transporter room, look straight ahead and before you could say 'Beam me up, Scotty' their bodies would dematerialise into tiny pieces and rematerialise, almost instantaneously, somewhere else. Though usually on a planet with a howling wind and polystyrene boulders.

While the programme's creator, Gene Roddenberry, used a number of scientific consultants for the show, the idea of a super-fast transportation device was born out of artistic necessity. Roddenberry needed something to get his space explorers on and off planets. Fast. Otherwise most episodes would be spent watching crew members read inflight magazines. More importantly, from a programme maker's point of view, the budget didn't stretch to the special effects needed to make a spacecraft repeatedly launch and land. Far better to remove the commute altogether, while at the same time turning it into a technological highlight. That way the crew could spend more time exploring new worlds and new civilisations, and Captain Kirk could spend more time snogging the female guest star of the week (Joan Collins's finest hour).

There is probably an entire generation, or two, that believes the idea of teleportation was invented by *Star Trek*, but this is not the case. It was a popular concept in science fiction from the 1930s onwards. A short story by George Langelaan, for instance, dealt with (unsuccessful) teleportation in the 1950s and was published in – of all things – *Playboy* magazine. This might sound obscure but the story was then made into the massively successful horror film *The Fly*. Several science fiction aficionados even claim that the first example of fictional teleportation occurs as far back as the nineteenth century, citing Edward Page Mitchell's 1877 story 'The Man Without a Body'. In it a man apparently (apparently because we've not yet tracked this story down to read it for ourselves) attempts to transmit a cat by wire in the same way that information is transmitted down a phone line. *Matrix* fans take note. And, rather like the eponymous fly, the cat doesn't fare too well in this form of teleportation either.

Despite lacking originality, there's no doubt that Roddenberry's teleportation device, the transporter, was an inspired invention. Teleportation would be the ideal, pain-free way to commute into work as no man has done before. It avoids traffic jams and train delays. It allows extra time to sleep in the morning and it's a lot safer than cycling during rush hour. OK, that's not strictly true. There was the odd mishap. Or, to use the correct *Star Trek* terminology, 'transporter malfunction'.

In an episode called 'The Enemy Within' (stardate 1672.1), a transporter malfunction produces two copies of Captain Kirk – one good and one bad. Both were halves of the same whole, each unable to survive without the other. The idea provided an interesting philosophical twist as well as giving Kirk the opportunity to snarl, sweat and grope the gorgeous Yeoman Janice Rand (the one whose blonde hair resembled a wicker basket). On occasion, the transporter also lost the deconstructed parts of a crewmember in outer space. Admittedly, it was never a character anyone had got to know (this is termed 'the red-shirt syndrome' by fans). But even if there was the odd disembodied interstellar scattering, let's look at the big picture here: if you want to avoid commuting, a transporter or teleportation device is definitely the answer.

A shame it's just science fiction? Well, prepare to put phasers on stun because in 1993 an international team of scientists published a paper in the journal *Physical Review Letters* claiming that teleportation was theoretically possible. It was not stardated 1 April, and the report's authors were respectable members of the scientific community. The effects were far-reaching. Their proposal was taken so seriously that it prompted teams of scientists across the world to take steps towards making *Star Trek* style teleportation a reality. It did not take long to get results.

In 1997 physicists from universities in Innsbruck and Rome created a worldwide outbreak of *Star Trek* puns by announcing the first successful teleportation. Shortly afterwards scientists from the California Institute of Technology, the University of Wales in Bangor and Aarhus University in Denmark reported teleportation from one end of a lab bench to another. Boldly going where no test tube had gone before.

In 2002, scientists from the Australian National University in Canberra reproduced the work and improved the technique further, producing yet more Trekkie headlines. More recently, in January 2003, teleportation was reported over distances ranging from 55 metres to 2 kilometres. Impressed?

You should be. Even if you've noticed we have not yet said exactly what these scientists teleported. That's because we first need to enter the scientific equivalent of the Twilight Zone: the world of quantum theory.

IT'S TRANSPORTATION, JIM, BUT NOT AS WE KNOW IT

Technically, all of the scientists mentioned above performed something called quantum teleportation. The key word here is 'quantum': the Latin word for 'amount' or 'how much', depending on the dictionary. In today's scientific parlance it means the smallest possible amount – or discrete unit – of energy or matter. Quantum teleportation describes a form of teleportation on a scale far smaller than anything the eye can see and at the very heart of matter itself.

The smallest particles of light are known as photons, and these can be described by their properties or quantum states. In the same way that a dress can have properties – of colour, style, fabric, etc. – a particle also has properties that provide us with more information about it. This could be which way a particle spins or the direction it vibrates as opposed to the length of its hemline.

The reason we're telling you this is because the first teleportation experiments were not performed with photons jumping from one part of a lab bench to another but with their properties or quantum states. It's the equivalent of transporting information about the colour or style of a dress but not the dress itself. Not quite on the level of zapping a carbon-based life form from an orbiting spacecraft to a planetary surface, but it's a start – and a remarkable achievement nevertheless.

Incidentally, you may have noticed that the definition of 'quantum' doesn't quite match the popular meaning of the phrase 'quantum leap'. This expression is usually used to describe an extremely large step or amount. Or, if you've seen the science-fiction series *Quantum Leap*, suspension of disbelief. How else can you accept the male actor Scott Bakula as a blonde woman in twin set and pearls?

If you can wipe those images from your mind (please) let's return to light and its smallest component, the photon. Before we can explain exactly how scientists achieved the first teleportation we need to take a brief trip to the start of the twentieth century. At this stage in our history scientists knew a lot about light. Or so they thought. More than 200 years earlier, that great apple-eater Sir Isaac Newton had used a glass prism to demonstrate that white light was composed of a spectrum of colours: red, orange, yellow ... Where are we? ... Oh, yes ... Richard of York Gave Battle In Vain ... green, blue, indigo, violet. We see these seven colours in a rainbow because raindrops have split light into its component parts.

There were several theories around as to the nature of light. Scientists tended to be divided into those, like Christiaan Huygens and Robert Hooke, who thought light was a wave and those, like Newton, who believed light was made up of particles. In 1801 an experiment by the English polymath Thomas Young convinced most people to choose one side of the argument and one only: that light must be a wave, not a particle.

Thomas Young was quite a guy. He could read at the age of two, was able to speak a dozen languages in his teens and, as well as being a physicist and a physician, helped Egyptologists decipher the Rosetta stone so that the modern world could understand hieroglyphics. Respect.

Young is perhaps best known, however, for the following experiment. He shone a light beam through two parallel narrow slits and produced what is now called a characteristic 'interference' pattern, consisting of light and dark stripes on a screen. Apparently, it took a couple of ducks to help him understand why light behaved the way it did. Young was supposedly watching some ducks swimming when he noticed that when the ripples from each bird overlapped they produced either a larger ripple or a patch of calm water. He then applied the same concept to light by imagining light travelling through the air in waves like ripples in a duck pond.

When the peaks of two light waves coincide, as in Young's double slit experiment, they combine to make a greater peak. This is called constructive interference. The combined light wave now has more energy and it produces a brighter spot. When opposing peaks cancel each other out, this results in no

A light snack

Light is fast. It travels in straight lines at 186,000 miles per second (approximately 300,000 kilometres per second) and this speed cannot be exceeded. It's the law and, unlike those on the roads, is always obeyed. Light travels at slower speeds through different mediums – such as water or glass – but at full throttle, through the vacuum of space for instance, that is how fast it is travelling, which gives you an idea of how far the Sun is from the Earth. Even at 186,000 miles per second, it still takes sunlight around eight minutes to reach us.

Our nearest neighbour, the moon, shows how light can be reflected. We only see the moon at night when it is in the right position to reflect sunlight (this is why we see it in different quarters too). Light can also be bent or refracted – examine a tilted pencil in a glass of water and you'll see what we mean.

In the same way that the colours of the rainbow are part of the light spectrum, light itself is part of an electromagnetic spectrum which includes radio waves, gamma rays, infrared, ultraviolet, microwave and X-rays. Our eyes are only sensitive to the visible part of the electromagnetic spectrum, which explains why we can't see infrared light or any other electromagnetic wave. The Scottish scientist James Clerk Maxwell discovered that light is an electromagnetic wave consisting of an electric and magnetic field oscillating at right angles to each other.

Picturing light as a wave is not too difficult. Watch a boat at sea and you'll see it bob up and down with the passing of each wave. A light wave is exactly the same, being made up of peaks and troughs. The distance from peak to peak or trough to trough of successive waves is called its wavelength and the frequency of a wave is the number of wavelengths per second.

The reason white light splits through a prism into red, orange, yellow, green, blue, indigo and violet is because each colour has a different wavelength. The wavelength varies depending on the colour's position in the spectrum. So red light, for instance, has a longer wavelength than blue light – 700 compared to 400 nanometres – a nanometre (nm) being one billionth, or 0.000000001, of a metre.

> Even though light behaves like a wave, it also consists of small particles or photons. This is the particle or corpuscular theory. In some experiments light will behave as if it were made up of particles while in others it will act as if it were waves. Or, as the British scientist William Lawrence Bragg put it, 'Physicists use the wave theory on Mondays, Wednesdays and Fridays, and the particle theory on Tuesdays, Thursdays and Saturdays.' We don't know how light behaves on a Sunday. This is-it-a-particle-is-it-a-wave? property is called wave–particle duality and, although it implies that light is a touch contrary, scientists seem to have come to terms with it.

energy being produced and so leaves darkness. This is known as destructive interference. There seemed no other possible explanation for such a characteristic pattern of bright and dark stripes on a screen. Light had to be a wave.

HOT BODIES

The first hints that light had other properties came when German physicist Max Planck first introduced the concept of units of energy – called quanta – in 1900. He invented quanta to help explain something that, in theory, wasn't supposed to happen.

Heat something and the colour of its emitted radiation changes with the temperature. This is why a poker in a fire glows red, then orange, then blue as it gets hotter. When scientists plotted graphs showing the intensity of radiation emitted by a hot body as a function of its wavelength they always got a similar shape: a curve that starts with minimum intensity and then hits a characteristic peak before trailing off again. The position of the peak can move too, depending on the temperature of the hot body.

We know that the temperature of the Sun is approximately 5,800 degrees C and the peak of its radiation curve at this temperature is in the yellow part of the visible spectrum, which explains why the Sun appears yellow. If the Sun were to get hotter, its radiation peak would move to a shorter wavelength, which is why that poker in the fire changes colour from red to orange to blue as it heats up. The problem for scientists in the past was that classical physics didn't predict this at all.

A hypothetical hot body that absorbs all the radiation falling on it was thought in classical physics to radiate energy (known as thermal or blackbody radiation) at an infinite rate and, according to the scientific laws of the time, equally at all wavelengths. The classical curve for the same graph therefore started infinitely high, with its maximum radiation always in the same region of the light spectrum: the ultraviolet wavelengths. This prediction bore no relation to reality: a poker in the fire doesn't emit ultraviolet rays when it gets extremely hot, otherwise we'd all get badly sunburnt. The incompatibility of theory with fact puzzled scientists so much that the anomaly became known as 'the ultraviolet catastrophe'. A touch over-dramatic, sure, but we are talking about something that threatened the centuries-old, classical understanding of how things worked.

The German physicist Max Planck was trying to explain the ultraviolet catastrophe mystery mathematically and found that his sums gave the right answer if he assumed that radiation was made up of small, discrete amounts of energy, or quanta (plural of quantum). In other words, energy was not smooth and continuous, as believed at the time, but quantised into small packets or chunks. These amounts would always be a multiple of a quantum and never something in between, in much the same way that you can only get multiples of £10 notes out of a cashpoint machine. If you can find one that isn't out of service.

It was a radical solution and one that Planck himself was never comfortable with, even when his ideas were eventually adopted by other scientists. Planck was reluctant to believe that quanta were anything other than a theoretical mathematical tool, but his formula worked and the mathematical constant it required to work became forever known as Planck's constant (known by the letter 'h' as presumably 'p' was already taken).

Five years later, in 1905, a fledgling scientist called Albert Einstein found that the same idea of quanta could explain the photoelectric effect. This was another scientific puzzle that couldn't be explained by classical physics. The fact that the solution used the idea of quanta added further credibility to

The classics

Classical physics is often referred to as Newtonian physics, after the English scientist Sir Isaac Newton (1642–1727), but, to be fair, classical physics really began a generation earlier, with the work of another eminent scientist, the Italian Galileo Galilei (1564–1642), who, as all Queen fans know, was magnifico-o-o-o. It must also include the work of James Clerk Maxwell on electromagnetic radiation. So, in general terms, classical physics can also be stated as pre-twentieth-century science – before quantum theory arrived to shake things up a bit.

This doesn't mean that all the previous science no longer applies. It just can't explain what happens at a quantum level. For larger-scale or macroscopic science Newton and his buddies still pretty much rule. Without Newton's laws of motion and gravity, for instance, man would never have gone to the moon.

Newton's first law of motion states that an object will remain at rest or in a state of uniform motion unless acted upon by a force. In practice this can rarely be seen because even if you set a marble rolling across an icy surface it will eventually stop because of the external force of air resistance, or drag. Newton's second law says that if you apply a force on a body it imparts acceleration (rate of change of its momentum) directly proportional to that force and in the same direction. His third law says that for every action there is an equal and opposite force or reaction. This is easily understood if you watch a game of snooker or hit a bouncer outside a nightclub.

As for that fruit-related incident? Who knows if an apple really did hit Newton on the head and produce a penny-dropping moment, but he did come up with a law of gravity. This states that there is a force of gravity between any two objects and that this force is related to their mass and the distance between them. There's an equation for this which, complete with a gravitational constant, is still used today to work out where to send spacecraft.

Planck's work. Yet Einstein, too, could never bring himself to accept all the implications of quantum theory.

If you shine a certain type of light on particular metals, such as zinc, in a vacuum then negatively charged particles called electrons escape from the metal surface and an electric current is generated. It's called the photoelectric effect (photo for light, electric for ... well, you can guess that one).

Heinrick Hertz and Aleksandr Stoletov both discovered this phenomenon at the end of the nineteenth century, but at the time no one knew how to explain what was happening. The effect didn't seem to happen with white light – no matter how bright – whereas classical physics predicted that if you increased the brightness, or energy, of a light wave, it should increase the energy of the electrons and cause more electrons to bounce off.

If that wasn't puzzling enough, electrons would only be ejected if the light hitting the metal surface was above a certain threshold frequency, usually in the ultraviolet range. Plus, if you made the ultraviolet light brighter you got more electrons but they weren't any faster or more energetic than when you used dimmer ultraviolet light. Confused? Don't worry. So was everyone else.

Luckily, Albert Einstein came to the rescue and provided a solution. By taking Planck's idea of quanta he considered light as individual particles, later to be known as photons. If a light particle hits a metal, the light transfers its energy onto an electron and this electron absorbs the energy. If this energy is enough to overcome the potential energy binding an electron to the metal, then off it goes. An electron can only receive energy from one photon, which explains why increasing the brightness of the light would make no difference, as if it's not at the right frequency to begin with, it ain't gonna happen.

The explanation of the photoelectric effect and his contribution to theoretical physics won Albert Einstein the Nobel Prize in 1921 and not, as might be expected, his work on relativity – something to remember for your next pub quiz.

While Planck's introduction of the quanta didn't signal the end of classical physics, it did show that classical physics was beginning to lose its grip as the fount of all scientific knowledge. It couldn't be applied to explain everything, particularly not how things behaved at an atomic level. As a result, a new branch of physics was born that required a completely new way of thinking:

quantum mechanics. Previous rules no longer applied, and some of the predictions of quantum theory would range from the strange and spooky to the downright unbelievable. But bear with us, as some of the strangest predictions are at the heart of teleportation.

THE MIGHTY ATOM

Quantum physics applies on an entirely different scale. Think of the 1966 film *Fantastic Voyage*, in which Raquel Welch memorably squeezes herself into a skintight jumpsuit. Now remember the plot. A team of scientists is miniaturised inside an experimental submarine so that they can journey through the human body to repair a blood clot in a scientist's brain. Obviously, this was before microsurgery but the whole premise of the film deals with life on a microscopic scale. Now think of something even smaller. And then some. We're talking about a world in which you can see not just individual atoms but everything that makes up those atoms.

It's easy to forget that our knowledge of the atom has only been forged in the last century. Its name derives from the ancient Greek *atomon*, meaning 'that which cannot be divided', because an atom was considered to be the smallest building block of matter, a fundamental particle that made up everything around us. It wasn't until 1909 that scientists discovered that the atom itself was made of something even smaller.

The revelation took place as a result of an experiment devised by the New Zealand scientist Ernest Rutherford to examine the structure of an atom. Alpha particles (ionised, or positively charged, helium atoms) were fired at an extremely thin piece of gold foil, thin enough to allow them to pass straight through. Rutherford believed that most of them would do exactly that, but he also thought that the recently discovered electron was part of an atom and that, if this was the case, some alpha particles would bump into these electrons and be deflected.

What took place stunned everyone. Most of the particles, as predicted, went straight through the foil. Some were also deflected but – and this was the stunner – every once in a while a particle bounced back. It was, Rutherford said, 'almost as incredible as if you fired a 15-inch shell at a piece of tissue paper and it came back and hit you'.

Two years later, and entirely from this unexpected result, Rutherford proposed a new model for the atom: that it was mainly empty space with its mass concentrated at the centre in a nucleus. A central mass would explain why – once in a while – the atom acted as a combination brick wall and trampoline to an incoming alpha particle.

The nucleus was surrounded by orbiting electrons in the same way that planets orbited the Sun – leading to Rutherford's idea often being called the planetary model. He used classical physics to form this atomic model and it was an important milestone. Even so, it only lasted a couple of years because, in 1913, another scientist put his theoretical oar in.

A GREAT DANE

Danish scientist Niels Bohr decided to incorporate quantum theory and Planck's idea of quantised energy into a new atomic model. He proposed the idea of electrons moving in circular orbits with each electron having a rate of rotation (angular momentum) related to a specific energy level. These electrons needed to absorb or emit an exact amount of energy – no more, no less – to break free from one orbit and jump into another.

If an electron leapt from one orbit to a lower energy orbit it would emit a photon with an energy equal to the difference in energy levels of each orbit. If an electron absorbed a photon that had the amount of energy needed to jump up a level, then it would do so. This model has since been revised to make way for newer theories, but the idea of quantised energy levels remains.

Bohr himself went on to be one of the main players in quantum theory. He was one of the scientists responsible for the so-called Copenhagen Interpretation of quantum mechanics and got into impressive scientific arguments with Einstein. Their intellectual ping-pong – Bohr always pro quantum theory, Einstein invariably trying to pick holes in it – furthered understanding of the whole field. It was Bohr who said: 'Anyone who is not dizzy after his first acquaintance with the quantum of action has not understood a word' – after a group of philosophers heard him lecture and immediately accepted this bizarre new concept.

Bohr also had an institute named after him at the University of Copenhagen that was financed by the Carlsberg Foundation. Probably the best institute of theoretical physics in the world.

HADRONS AND LEPTONS AND QUARKS, OH MY!

Today's understanding of the atom is even more detailed. An atom always looks pretty crowded in school science books, but most of an atom is, in fact, empty. The nucleus is about 10,000 times smaller than the atom itself.

Inside the nucleus are positively charged particles called protons and uncharged neutral particles called neutrons. But outside the nucleus is a whirl of activity: tiny negatively charged particles called electrons, much smaller than the nucleus, and in a state of constant motion. The best picture of an atom is therefore one in which electrons are represented by a hazy cloud around the nucleus, as this shows where the electrons are most likely to be found. This cloud can be different shapes too, so it's a much more complicated picture than the popular emblem of many a science-fiction film, in which electrons orbit the nucleus in fixed orbits like small planets around a Sun.

The idea of an atom being indivisible has also long been shattered. Electrons remain 'fundamental' or indivisible, and they belong to a family of particles called leptons, which are also fundamental. But scientists now know that protons and neutrons belong to a family called hadrons, and these types of particles are made up of even smaller particles called quarks. There are six types of quark and they are sociable creatures because they are always found in pairs. Quarks exist in three main flavours, or types: up and down; top and bottom; strange and charm (we are not making this up). The sub-atomic world is a truly curious place and the word quark itself has a quirky origin. It's a nonsense word invented by James Joyce in his novel *Finnegan's Wake*.

WHEN TWO BECOME ONE

A century ago, bizarrely named sub-atomic particles were unknown to scientists but their existence can be traced back to that one single moment in 1900 when Planck introduced the idea of quanta. Einstein took the next crucial step with his explanation of the photoelectric effect and his light quanta (later to become known as a photon). Then, when French scientist and aristocrat Louis Victor de Broglie decided to apply quantum theory to Bohr's model of the atom, the nature of light was made clearer still.

Bohr put forward the idea that light was both a particle *and* a wave and that all matter had the same wave–particle duality. This meant that particles could behave like waves and waves could behave like particles. De Broglie then came up with a formula that linked a particle to a wave. This equation is a surprisingly simple one, as the best ones often are, and it related the momentum of a particle (its mass and velocity) with the wavelength of the wave associated with that particle.

Three years after this proposal, an experiment by the American physicist Arthur Compton confirmed that one particular particle, the electron, did in fact have wave-like properties and became proof that electromagnetic radiation had both the properties of a wave and a particle. So Isaac Newton and Thomas Young were both right. Light is a particle and a wave.

AN IMPROBABLE SCIENCE

In 1926 an Austrian scientist, Erwin Schrödinger, provided the mathematical equivalent of the other point of view put forward by Bohr: how to describe a particle as a wave. He formulated an equation explaining the type of wave that describes the movement of a particle – any particle – including a photon. This is known as the particle's wave function and it encapsulates the particle's quantum state. It is also a mathematical way of giving scientists information about the particle and, as there will always be some uncertainty involved, this description includes probability. Probability because, according to quantum theory, you cannot say for sure exactly where a particle will be – only where it is likely to be.

It's rather like trying to say exactly where your partner is once they've left the house in the car on the way to the cinema. An hour later there's a distinct probability that your loved one is inside the local flea-pit watching a film. Having said that, there's also a probability that he or she may have popped into the pub for a drink, met a friend and never reached the cinema. There's also a probability that the car broke down en route and your partner is on a street somewhere under the bonnet. Or, worst of all, your other half is just around the corner having an affair with your best friend.

The truth is, unless you know for sure, once that person has left your sight there exists a range of probabilities for his or her whereabouts. You think you know where someone is but in reality until you call them on a mobile phone and verify their position with a global positioning satellite, you don't know with 100 per cent certainty where they are. Until then all probabilities are possible. Now apply that suspicious way of thinking to a particle. It means that until you locate that particle or determine its particular property or quantum state through measurement, there exists a range of probabilities corresponding to where that particle could be or what properties that particle might have.

Schrödinger's wave equation highlighted two extraordinary aspects of quantum theory. The first was this notion of uncertainty – an idea that shocked many scientists, Einstein among them. This was not the reason for his 'mad scientist' hair – that look came later – but it did produce his famous 'God is not a gambler' quote. 'Quantum mechanics is certainly imposing,' he said, 'but an inner voice tells me that it is not yet the real thing. The theory says a lot, but does not really bring us closer to the secret of the "Old One". I, at any rate, am convinced that He is not playing at dice.'

Schrödinger's equation also led to an unusual condition known as super-position, in which particles can be in different states at the same time. This arises mainly through mathematics and something called Fourier analysis, which shows how one wave function (the description of a particle as a wave) can be seen as the sum of different wave functions. So, on a simplistic level, if you have one big wave it could be seen as the overlapping sum of many smaller waves. In this case any particle in that wave could have numerous positions and properties. It's not until you make a measurement that one distinct possibility is chosen. Another way of understanding this is to imagine you have a quantum sweet bag filled with assorted pick 'n' mix. Once you put your hand in the bag and pull out a sweet, all the other sweets are suddenly gone.

At around the same time that Schrödinger formulated his ideas, a German physicist called Werner Heisenberg came up with similar mathematical conclusions. In 1927, Heisenberg's Uncertainty Principle showed that it was impossible to determine both a particle's speed and its position at the same time. The reasoning behind this is that if you get too close to the particle you are trying to measure, you influence it and affect its behaviour. The particle will then be in a different position from where it originally was, or, because of your interaction, travelling at a different speed.

This will cause us big problems if we want to build a transporter, because we need to know all the information about particles and atoms inside a human body in order to get the same rematerialised collection at the other end. Just ask actor Jeff Goldblum, who appeared in the remake of *The Fly*. This superior version graphically showed what happens when you try to build a transporter device and don't clean it thoroughly before use. The transporter didn't receive all the essential teleportation information about the human body – because it got mixed up with that of a house fly – and who can forget the result?

The technical advisers on *Star Trek* found a clever way to solve the problem of uncertainty for their teleportations by introducing a Heisenberg compensator. This could, surprise, surprise, overcome Heisenberg's uncertainty principle. As we're talking science fiction here, no episode has ever gone into how it works (what a dull episode that would be), which is why, when asked how it worked, *Star Trek* scriptwriter Mike Okuda replied 'Very well, thank you.'

NOW YOU SEE IT, NOW YOU DON'T

The introduction of probability, uncertainty and the ability of something to be in many places at the same time (known as multiple superposition of states)

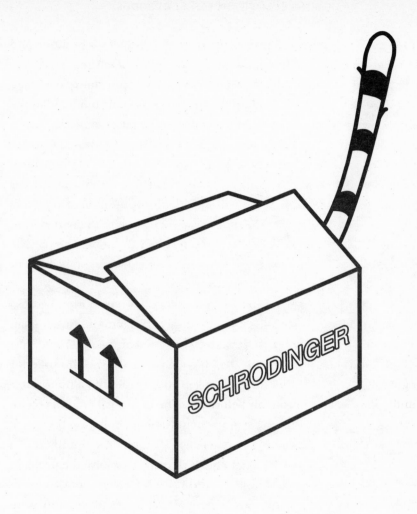

does give quantum theory an air of magic and science fiction about it. So much so that you would have thought the theory would have been burnt at the scientific stake a long time ago. But quantum theory has been experimentally tested and, so far, it works. Understanding *why* quantum mechanics behaves the way it does is another matter. Take heart from popular physicist Richard Feynman, who once said, 'I think it is safe to say that no one understands

Schrödinger's cat

Erwin Schrödinger, it is fair to say, was probably more of a dog person. Why else would the Austrian physicist imagine such an inventive way to kill a cat? Correction. Maybe kill a cat.

Schrödinger's idea was to show how ludicrous the new quantum physics could be. He did this by thinking up a Dr Phibes type 'diabolical device': a container of poisonous gas that is only released if radioactive material starts to decay and gives off radiation that will trigger the container to open. This potential cat-killing gas is inside a windowless box with a cat (type not specified) and all that the cat needs to survive.

The cat's life depends on whether or not an atom is released from the radioactive material. Quantum theory states that if you don't observe what's going on there is a range of probable outcomes. There was nothing in the English language to describe this possibility so scientists called it a superposition of two possibilities or a superposition of states. This means that either the atom has been released, or it has not been released or it both has and has not been released. As the cat's life depends on this atom, Schrödinger extended this concept to the cat too: if the box is opened there is a 50:50 chance that the cat will be dead and a 50:50 chance that it will be alive, but that if the box remains shut the cat's state is undetermined. It is both alive and dead.

Obviously a cat would never be both alive and dead in real life and this is why Schrödinger introduced this thought experiment. He believed the implications were so patently absurd that it would demonstrate the logical flaws of this new science.

quantum mechanics. Do not keep saying to yourself, if you can possibly avoid it, "But how can it possibly be like that?" because you will go down the drain into a blind alley from which nobody has yet escaped. Nobody knows how it can be like that.'

With that quote in mind, please just accept that in quantum theory a particle has no definite properties until they are measured. It's rather like tossing a coin. You cannot say whether the coin will land heads or tails until you slap a hand over it. Now if that coin was smaller than an atom, then while it is spinning and you don't know the outcome of the heads/tails scenario, the coin would be in a superposition of states. The probabilities then are that the coin is heads, tails or both heads and tails.

Once a measurement is taken (a slap of the hand) then we will know for sure, as one particular set of properties will appear, for example heads, and all the other possibilities will disappear. The weird bit is that until the slap all the other options exist and the coin can have the entire range of properties available – the magical superposition of states. This ability, as we will see, comes in useful for making teleportation work.

If you still think it sounds crazy, you are not alone. Bohr got it right when he said, 'Anyone who is not shocked by quantum mechanics has not fully understood it.' No wonder Schrödinger devised his now famous thought experiment (one done inside your head rather than inside a laboratory) to show exactly where these bizarre ideas would lead.

SPOOKY SCIENCE

Schrödinger wasn't the only scientist to point out the seemingly illogical results of applying quantum theory to particles. In 1935 Albert Einstein also wanted to show that quantum mechanics was flawed. To do so he wrote a paper with Boris Podolsky and Nathan Rosen, attempting to poke holes in the theory. The key to the authors' discomfort is in the title: *Can Quantum-Mechanical Description of Physical Reality Be Considered Complete?* In their opinion the answer was definitely no. They set out to convince other scientists likewise.

Einstein, Podolsky and Rosen proposed a thought experiment based on quantum theory that, in their opinion, proved that quantum theory was incorrect. They argued that if you separated two particles described by the same wave function then you could make individual measurements on each particle without disturbing the other. Under these conditions, knowing something about one particle would automatically tell you something about the second particle.

When a particle is moving, for instance, it has a property known as momentum and this momentum depends on its mass and velocity. A stationary particle has no momentum (because it is not moving) but if a moving particle bumps into it then a transfer takes place. Imagine that these particles are snooker balls and you can see what we mean. A moving white ball transfers some of its momentum to the stationary black ball. One ball slows down. The other starts moving, hopefully into the corner pocket. After the collision the total amount of momentum shared between the two balls will equal what the white ball had before the collision. This is the principle of conservation of momentum. It's the same with particles. If you know the momentum for both particles, separate them and then measure one particle's momentum, you can do a quick bit of maths to work out the momentum for the other particle.

In other words, measuring something about one particle in one place would give you knowledge of the other, separated particle in another place. These two particles would be somehow connected. This went against common sense and, as this phenomenon didn't happen in the real world, it became known as the EPR paradox. Einstein and his colleagues concluded that the theoretical existence of such a connection meant that the theory was incomplete and there must be 'hidden variables' for it to all make sense.

The paper provided a mathematical description of two particles linked by specific properties: position and momentum. This connected or 'correlated' condition was later to become known as an entangled state, or EPR beam or pair, after the initials of Einstein, Podolsky and Rosen. An irony, all things considered, because despite their assertions that entanglement couldn't happen, not only was it later shown to exist but it became the basis of the first teleportation experiments.

Let's not be too harsh. The reason these scientists missed the incredible possibility of teleportation was understandable. They simply made an obvious but incorrect assumption of locality: that what you did to a particle in one location would have no affect on another particle in another location. And who wouldn't think that? No wonder Einstein called it 'spooky action at a distance'.

In the 1950s an American scientist, David Bohm, wrote another, simpler version of Einstein's paper using the state of spin rather than position and momentum. Then, in 1964, an Irish physicist called John Bell made an important breakthrough about the so-called EPR paradox: it wasn't a paradox after all.

Einstein and co had assumed that quantum theory must be wrong because otherwise you would have a bizarre situation in which what you did to one particle influenced another in a different location – non-locality, as it is known in the field. But Bell thought exactly the opposite, and he showed mathematically, through what is now known as Bell's theorem, that non-locality could happen and quantum theory was right. No one took much notice until the 1980s, when French scientist Alain Aspect proved experimentally that when it came to these special correlated or connected particles the left hand really did know what the right hand was doing. Entanglement was real.

ENTANGLEMENT

Entanglement is a mind-boggling concept; it is also not an easy thing to do. But there are several ways to produce entangled particles. One method is called an atomic cascade and involves exciting an atom. This is not as kinky as it sounds. It's a scientific term for boosting energy and can be done by using ultraviolet radiation or a laser. The electrons in the atoms gain extra energy and, if the amounts are right, they can jump up two energy levels. When they drop down again from these levels two photons are emitted and these two photons are entangled. This doesn't happen very often so there's a cause for celebration when you get an entangled pair.

Entangled pairs can also be produced when an electron smashes into a positron – its antimatter equivalent – and they annihilate each other. They also result from shining a laser into special types of 'non-linear' crystals. These are crystals that do weird things with light. When a laser passes through a non-linear crystal sometimes a photon is split into two. The two smaller photons have half the frequency of the original photon and are of lower energy. Only one in 10 billion ultraviolet photons produce a pair but the most important thing is that the two photons are entangled.

Jeff Kimble at the California Institute of Technology has provided one of the best definitions of entanglement. This is his take on the connection between two entangled particles: 'Entanglement means if you tickle one, the other one laughs.' In theory an entangled pair of particles could be created and sent separately into space. If one particle's properties were then measured the other particle should instantly be affected with complementary properties. Yet if this happened when the particles were millions of miles apart, the instructions for this to happen would have to have travelled faster than the speed of light and, as Einstein had decreed years earlier in his Special Theory of Relativity, nothing travels faster than the speed of light. Entanglement, the reasoning went, must not be able to happen or it would break the laws of physics.

SCIENCE FICTION TURNS FACT

Not surprisingly, when faced with the prospect of quantum information being unable to exceed light speed or overcome the uncertainty principle (as pointed out in the EPR paper), dreams of teleportation seemed pure science fiction. Thankfully for those of us who are hoping for an end to commuting, this all changed in 1993.

Six international scientists – Charles Bennett at IBM, Gilles Brassard and Richard Jozsa from the Université de Montreal, Claude Crepeau of the Ecole Normale Supérieure in Paris, Asher Peres of Technion in Israel and William Wooters of Williams College, Massachusetts – published a paper that described a way to achieve quantum teleportation without violating any scientific laws. It relied on the principle of entanglement.

The team suggested a way of teleporting quantum states of a particle with entangled or EPR beams by using an essential feature of quantum information – that it can be swapped from one system to another but never duplicated or cloned. Their paper appeared in the *Physical Review Letters*, the same journal that had published the EPR paper almost 60 years earlier.

Imagine someone called Alice who wanted to send her mate Bob some information about the quantum state of a particle. Don't ask us why, just go with the flow. The six scientists showed, on paper, how Alice could send this information to Bob by using the unusual abilities of two entangled particles. Provided Alice has one entangled particle and Bob has the other – the one that shares its quantum telepathy – teleportation can take place.

This is why. If Alice tries to measure something about her particle and then shares this information with Bob via a conventional communication channel – by fax or email for instance – Bob will only ever receive an imperfect copy of her original particle. This is because the uncertainty principle will not allow every piece of information about that particle to be measured. Plus, simply by making a measurement, Alice will have altered the state of her particle. As a result Bob's copy will always be inaccurate. If it weren't, the procedure would violate both the uncertainty principle and the no-cloning theorem. And that's bad.

For teleportation to work, Alice and Bob must each have two halves of an entangled pair, one particle each, and take full advantage of their spooky connection (the left hand knowing what the right hand is doing). If Alice then deliberately entangles her particle – the one whose property is to be teleported – with the previously entangled particle, she can no longer tell which is which. This is important because any measurement taken of both these particles mixed together will not ruin things and alter the state of her original particle.

Alice's particles now share something unique: quantum entanglement – this ability to connect. Alice's entangled particle will therefore complement any specific property of her original particle. But let's not forget that Bob has a particle that used to be entangled with one of Alice's entangled pair. So his particle, despite being in another location, will automatically complement that particle too, and in doing so will end up with the same property as the particle to be teleported.

For a system with two states (say, spin up or down; heads or tails if we think coins again) there are four possible forms of entanglement. The trick is in letting Bob know which the correct configuration is so that he gets the right teleported state at his end. This is done by Alice measuring her two particles and sending the result to Bob. This result tells Bob which of the four entangled states Alice created between her two particles and will specify how the original state will appear at Bob's end.

At this stage Bob's particle doesn't have the same quantum state as Alice's particle because when Alice makes her measurement her two particles are entangled. Bob's particle is therefore a contorted version of the quantum state Alice wanted to send him. But once Bob is told how Alice's particles are entangled, or correlated, with each other he knows enough information to 'untwist' his particle into the state of the particle Alice wanted to teleport. It's almost as if his particle has suddenly become a mixed-up Rubik's cube of information and Alice's classically sent message, together with quantum telepathy, tells him how to straighten it out.

When Bob has both these pieces of information he can make the relevant adjustments to produce an accurate replica of the original particle's state. At the same time, at Alice's end, the original particle no longer exists in its original form. To all intents and purposes it has disappeared. Sound familiar? It did to the scientists too, so the team borrowed a well-loved science-fiction term to describe what was happening: teleportation. The scientists were quick to stress, however, that, unlike most sci-fi teleportation, no scientific laws were being broken using their method. So there.

There are a few important things to take notice of here. Alice never knew all of her particle's quantum states to begin with, so the uncertainty principle remains intact. Also, the original is destroyed, so no clones are produced.

Finally, because part of the information had to be sent classically, the teleportation process is not instantaneous and would not exceed the speed of light. Something that would certainly please engineer Scotty in *Star Trek*'s transporter room. 'You cannae break the laws of physics.'

Despite using a science-fiction term and adhering to all the known laws of science, the paper pointed out that this type of teleportation could do something science fiction could never do: the sender doesn't need to know the receiver's location. It wouldn't matter where Bob was – he could be on the end of a phone line and not tell Alice which country he was in – teleportation would still happen. In *Star Trek*, teleportation could never take place without the transporter operator knowing the co-ordinates of the teleportee. It all added to the drama.

The original concept also produced rather an endearing convention among other scientists working in this field. From that time on the sender of the particle to be teleported has always been referred to as Alice and the receiver as Bob.

TWENTIETH-CENTURY TELEPORTATION

The paper by Bennett, Brassard, Jozsa, Crepeau, Peres and Wooters was a spectacular piece of theoretical physics. It is a measure of the quality of the individual scientists' work and reputations that their theoretical seal of approval for teleportation wasn't instantly dismissed as fantasy. Instead, scientists all over the world set to work out how to do it.

It didn't take long. In December 1997 two research groups from the University of Innsbruck in Austria (led by Anton Zeilinger) and the University of Rome in Italy (headed by Francesco de Martini) reported the first stages of teleportation. They teleported a particular quantum state of a light particle or photon: polarisation.

If you own a pair of Polaroid sunglasses then you can see an immediate demonstration of how light can be polarised. Simply hold your sunglasses up to the window and rotate them. Unless you've been shopping in a Bangkok street market and have been sold a dud, the amount of light should change because that's what sunglasses do: they block light travelling in a certain direction. This is because light waves can be horizontal (like a snake wiggling from side to side across a desert), or vertical (imagine picking up aforementioned snake by the tail and wiggling it up and down), or at any angle in between (just go crazy with that snake).

This is how the Innsbruck team (Dik Bouwmeester, Jian-Wei Pan, Klaus Mattle, Manfred Eibi, Harald Weinfurther and Anton Zeilinger) achieved their teleportation. A photon was sent from a sending station called Alice, passed through a filter and given a 45-degree polarisation. Meanwhile the scientists had also created an entangled pair by passing a photon of ultraviolet radiation through a non-linear crystal. This produced two entangled lower energy photons – one of which is sent to Alice, while the other goes to Bob at a receiving station.

Alice's photon and one of the entangled photons are then sent towards a beam splitter so that they arrive there at the same time. A beam splitter is the fancy name for a half-silvered mirror which can either reflect or transmit the two photons with a 50:50 probability towards one of two detectors.

If these were ordinary photons then the probability of them each going to a different detector would be one in two. But these are no ordinary particles. As they've been entangled normal rules don't apply and quantum theory predicts a one in four probability instead. Sure enough, 25 per cent of the time the photons travel to two different detectors and teleportation takes place.

Alice's photon cannot be identified when it is entangled with one of the previously entangled photons, so any measurements she makes do not upset Heisenberg's uncertainty principle. The result of her measurement is sent to Bob classically but there's also a quantum route where the telepathy kicks in. All that's known of Alice's two photons is that they must have complementary polarisation (if one is horizontal, the other will be vertical, for instance). But the same spooky connection applies to the photon that was entangled with Bob's photon too (there's a lot of particle swapping going on here). The combination of Alice's conventionally sent message and the quantum tele-pathy leaves Bob's photon with the same polarisation as Alice's original photon – once a bit of Rubik's cube twisting has gone on at his end.

The no-cloning law of teleportation is not violated either. By losing its iden-tity during entanglement, the state of Alice's photon is destroyed during tele-portation. A lesson for anyone seeking entangled relationships perhaps. Bob must still do some tweaking at his end using the classically sent information from Alice, but the most important thing here is it worked. Scientists achieved the world's first teleportation.

The Innsbruck team teleported a photon from one side of a lab bench to another – a distance of about 1 metre. Technically speaking, they teleported a photon's quantum state – in this case its polarisation – although the origi-nal photon was destroyed and an exact replica was created in another loca-tion. Unfortunately, this replica had to be destroyed in order to verify the teleportation and make it work, which is a slight flaw for any future human transporters. But remember, we are talking about proof of concept here and that's what the Innsbruck team did. Their work also highlighted, as Bennett's paper had done several years beforehand, that entanglement could be an important ingredient for a new type of information transfer: quantum computing.

To say that quantum computing has great potential is an enormous under-statement. Millions of dollars have gone into funding its research around the world. Whoever gets there first will be able to crack any code, perform any calculation and have unimaginable computing power compared to what is available today.

Richard Feynman was one of the first to show how to do computations on a quantum scale in the early 1980s. Oxford University physicist David Deutsch then published a paper in 1985 showing that, in theory, a quantum computer could work and would outperform any traditional machine. The research began and, once some mathematical difficulties had been ironed out, the whole notion of quantum computers gained momentum.

Quantum computers

Computers process information by using bits. In a physical sense these bits can be magnetised regions on a disk or electrical charge stored in a capacitor. They can have two states – either on or off – represented by a 1 or 0. These are often referred to as binary bits because they are the only numbers in a binary or base 2 system (compared to the base 10 decimal system, which uses numbers 0–9). In computer terminology eight of these bits make up a byte and there are four bits in a nibble. Strange but true.

A quantum computer would, like everything related to quantum physics, be on an atomic level and so would need its own type of quantum building block, such as the spin of an electron. As this type of quantum bit is on a totally different scale from the bits used by today's computers, it has been named the qubit.

Bits are good but qubits are potentially much better and here's where qubits have the edge. Qubits, because of their quantum nature, can have more than the two states of a conventional computer, because they can be 0, 1 or (like Schrödinger's cat) both 0 and 1. This means that if you have a pair of qubits there are four potential states – 00 (both off), 11 (both on), 01 (one off, one on) and 10 (one on, one off). If you add more qubits, the number of states rises dramatically because it is an exponential rise. Just ten qubits would therefore give two to the power of ten, or 1,024 states, the equivalent of a kilobit in conventional computing.

This superposition of states means that a quantum computer consisting of hundreds of qubits could work on all the possible outcomes of an extremely complex problem simultaneously, as if it had billions of separate modern-day processors. This would produce, in effect, a super-fast computer of almost unimaginable power compared to the ones we have today. And don't information scientists know it. If obtainable, quantum computing would not only revolutionise the industry, it would also solve a potential stumbling block for existing computer technology.

One of the founders of Intel, Gordon Moore, predicted that the number of transistors in computer chips would double every 18 months. This has become known as Moore's Law and still holds as chips continue to get smaller and computers faster. Miniaturisation, however, can only go so far because once you reach the sub-atomic level then quantum mechanics kicks in and upsets all the rules. Quantum computing could be the answer.

This is definitely an exciting time for scientists working with quantum information systems. In 1998 scientists at the Los Alamos National Laboratory in the United States built the first three-qubit quantum computer, capable of performing simple maths. The same laboratory built a seven-qubit quantum computer in 2000 in a drop of liquid. It will be years before a complex quantum computer is realised, but progress is definitely being made.

ANY ADVANCE ON A PHOTON?

It's all very well using a theoretical Bob and Alice to transport photons, but what about teleporting Bob and Alice? A year after the Innsbruck success there was another important advance. It was reported in the journal *Nature* in 1998 by a team consisting of Akira Furusawa, Chris Fuchs and Jeff Kimble from the California Institute of Technology; Jens Sorensen and Eugene Polzik from Aarhus University in Denmark and Samuel Braunstein from the University of Wales in Bangor.

They performed quantum teleportation on hundreds of photons and with several quantum states using the same method of entanglement but with a few differences. They used two crystals – the first inside a cavity so that when a laser beam passed through the crystal it split each incoming photon into two photons with double the frequency. More photons resulted, producing a much stronger beam of light.

This new beam of light was then passed through a second crystal, but this one halved the frequencies, producing what has been labelled 'squeezed light'. True, they ended up with light of the same frequency they had in the beginning, but with one important difference: the beams were now entangled.

To the two crystals and the now familiar terminology of Bob and Alice, the

Quantum encryption

People have been sending coded messages for thousands of years. But every time cryptographers think they have an unbreakable method – such as the Enigma machine in the Second World War – someone (the Poles in the case of the Enigma machine) comes up with the answer. Modern codes bypass the code-breakers by being so complicated that the solution, although obtainable, would take years to calculate with even the most powerful of supercomputers. This is where quantum encryption is in a league of its own. It is one of the most promising spin-offs in the world of quantum mechanics.

Quantum encryption was the brainchild of Charles Bennett and Gilles Brassard (two of the team who originally proposed teleportation). The idea was to encrypt a message using the polarisation of photons (the different directions that a light particle can vibrate). If someone tries to intercept a quantum-coded message, the very act of interception forces the particles into a particular state and alters the message. It also lets the sender and receiver (Alice and Bob) know that their quantum line is being tapped. Most importantly of all, because of the quantum nature of this sort of encryption any quantum-coded message would be unbreakable.

Brassard and Bennett's theory was proven in 1988 when the first quantum code was sent between two computers 30 centimetres apart. In 1995 a team at the University of Geneva increased this distance to 23 kilometres using an optic fibre to send the photons. In January 2003 the University of Geneva also reported the first-long distance teleportation in the journal *Nature*. The scientists had teleported photons a distance of 2 kilometres (more than a mile) using the sort of glass fibre found in standard telecommunications cables. This was a significant development for those wanting to send encrypted messages, because photons can only travel so far under their own steam. This leads to the possibility of using teleportation as a repeater to copy the message and allow it to be sent along cables as far as you want the message to go. This isn't the only possible method, as Arthur Ekert suggested using entanglement for encryption and this too has been experimentally verified. But the field is moving so fast that some people are predicting that quantum cryptography could be used within five years. In June 2003, British scientists from Toshiba Research Europe Ltd in Cambridge demonstrated quantum cryptography over 100 kilometres of optical fibre. The race for unhackable communications is definitely on.

team introduced 'Victor the Verifier' and used a technique by which the tele-ported photon didn't have to be destroyed to prove that the experiment had worked – a definite improvement if we're looking at making that *Star Trek* transporter a reality.

In 2001, Eugene Polzik and his team at the University of Arhus in Denmark entangled trillions of atoms in two clouds of cesium. Human teleportation would require the entanglement of atoms on an even greater scale, so there's a long way to go yet, but there are more immediate uses of this emerging tech-nology. As well as quantum computing there's quantum encryption, a tech-nique that can make full use of the properties of entanglement. Peter Shor, who worked for AT&T Bell Laboratories in the United States, was one of the first to be interested in adapting quantum teleportation for encryption.

NOW WILL YOU BEAM ME UP?

It is the year 3050. After a quick pre-breakfast jog on a Sydney beach you take a sonic shower, put on a figure-hugging Lycra suit and prepare for your morn-ing commute to work. It's a tough one. Just over 10,000 miles in fact. But luck-ily a transporter reduces this journey to a couple of seconds – although somehow you still manage to be late; you always underestimate the number of people queuing up to undergo teleportation. Fortunately, there's the week-end to look forward to: a mini break taking in the Great Wall of China, the Taj Mahal and a polar-bear-watching trek in the Arctic Circle. We can dream? Sort of. There remains an enormous technological gap between teleporting particles of light and teleporting the number of atoms that make up the human body.

Ping Koy Lam, from the Australia National University, thinks human teleportation will never happen because, 'There are about 10,000,000,000,000,000,000,000,000,000 atoms in a human and at the moment no one has teleported an atom yet.' Fortunately for those fed up with commuting, not everyone is so pessimistic. Although it might take a while before anyone can benefit, some scientists do believe that human teleportation may be possible. Eventually. Teleportation expert Sam Braunstein, for example, reckons the prospect of beaming people rather than photons across a lab is feasible, but it's at least 100 years away.

If you're reading this book on the way to work in a crowded train or bus, with someone's elbows in your back, this may not be much consolation. But think how your great-great-great-grandchild's quality of life will be improved by never having to commute for more than a few seconds. More time for Tomb Raider version 3002 perhaps, or to take some flying car driving lessons. Teleportation would also take the haul out of long-haul holidays, although it may not mean the end of lost luggage. Your holiday clothes could still end up on the other side of the Earth; they'll just get to the wrong destination a whole lot faster.

There is, however, an important philosophical point to consider before you jump onto that transporter. Successful teleportation means exact duplication has taken place. This will not be cloning, by the way, as human clones (identical twins) are genetically identical yet not necessarily physically identical. In teleportation once you are rematerialised at another location you would be both genetically and physically identical to the 'you' that was left behind.

We already have technologies that can duplicate information, but quantum teleportation differs from a fax or an email, say, in one crucial aspect: once the duplication is completed the original is always destroyed. The original 'you' will be disassembled so that the exact copy becomes, to all intents and purposes, the new original. Every physical aspect of your body will be reproduced. From your fingerprints and that mole on your arm to each individual atom that makes up the human version of you.

It follows that the new original brain will also contain those years of memories, learning and experience, but what happens to your soul? This is assuming we all have one, of course. It's an important question to answer, otherwise how do we know who will arrive at your workplace? We may all go through the motions sometimes (particularly after a night out), but we need to be sure that it will be you at the office and not a soulless Stepford-wife-style copy who lacks your individual va va voom.

So how do we define our soul? Many people around the world believe we each have a soul and that it is connected to, yet at the same time separate from, our physical embodiment on Earth. The trouble is, no one has ever proved that the soul exists, and if the soul cannot be explained by physical means how do we know if it can be teleported? Maybe the soul doesn't need a device to keep tabs on its physical receptacle and will automatically rejoin the transported body. Maybe a teleportation machine could prove or disprove the existence of the soul. For that reason alone, the pursuit of a human teleportation device may be worth following.

Einstein once said, 'The more success the quantum theory has the sillier it looks.' And who can argue with that? Teleportation is real. It has been experimentally proven for quantum states of a photon, a light beam and a cloud of photons. The next stage is to teleport a whole molecule or a virus. Entanglement of particles, Einstein's 'spooky action at a distance', happens. What you do to one half of an entangled pair affects the other separated entangled partner. There is a long journey between where we are now and being able to step into a transporter and end the daily commute for ever. But human teleportation might, just might, be possible. One day.

CHAPTER FOUR

How to Lose Your Love Handles

It's said that inside every fat person there's a thin person trying to get out. Sometimes more than one. And it's easy to put on weight. Very easy. The occasional Mars Bar, the odd packet of crisps or dollop of ice cream (OK, tub) and before you can say elasticated trousers you're wheezing to catch the bus and struggling to fit through the doors once you get there.

Wouldn't it be great if science offered a solution whereby you could eat what you wanted, when you wanted, without piling on the pounds? Better still, if technology could tackle cellulite, improve muscle tone and make you once again look young, slim and beautiful. After all, wouldn't it be more rewarding to *attract* the perfect blonde (or brunette or redhead) rather than have to go through the whole palaver of cloning one?

Beauty may be in the eye of the beholder, but for those who've had a brush with the ugly stick, the number of beholders is severely restricted. So what if science could change you for ever, sculpting your body in the image of Brad or Jennifer, J-Lo or Jude Law? Could science really make your shallowest dreams come true?

With gene therapy and genetic modification we can rebuild you. We have the technology. It won't involve any major surgery just a little bit of risk. Well, actually quite a lot of risk, but we'll come to that later. For now, sit back, relax and munch on your favourite snack, as we explain the harm it's doing to your body.

Balancing act

When researching this chapter we came across a long out of print medical textbook. Inside, among some gruesome diagrams, there was a section on obesity. The gist was that people put on weight when energy input exceeds energy output. In other words, people get fat when they eat too much and don't take enough exercise. Although the text was written back in the 1930s, when it wasn't fashionable to blame being overweight on heavy bones or allergies, the message still holds true today. Eat too much and you get fat. But whereas 70 years ago obesity was the preserve of the privileged, today experts consider it a disease of epidemic proportions.

Most of us are probably carrying a few extra pounds and might be a tad or several tads overweight. What worries health professionals, and indeed the World Heath Organisation, is how many people are now clinically obese. The amount of excess body fat carried around can be measured using something called the Body Mass Index, or BMI. (Also the initials for a UK-based airline but we'll resist facile comments about jumbos. Maybe later.) The BMI is based on the ratio between your weight and height. People with a 'normal' weight have a BMI of between 18.5 and 25, people who are overweight have a BMI of between 25 and 30, and those who are considered obese have a BMI of 30 or greater. Obese: even the word itself sounds unattractive.

Recent figures from the International Obesity Taskforce estimate that up to 1.7 billion people in the world are overweight or obese. The biggest problem is in the United States, where getting fat is practically a constitutional right (see the box on page 111 for more on this). But if you're not American there's no cause for gloating; the rest of the world isn't far behind. At the moment almost half of all adults worldwide are overweight. Most of us could stand to lose a few pounds.

Staying alive

Obviously, the last thing any of us wants to be told is that we need to eat less and take more exercise. We want to be able to eat more *and* lead the increasingly

Fat nation

For land of the free read land of the fat. America is suffering from what doctors call an obesity epidemic. If you've ever had your body space invaded by a victim of fast-food culture on a US flight, you'll know exactly what we mean. In Mississippi, for instance, more than 62 per cent of the population are considered overweight, and 24 per cent are considered obese. Which doesn't leave many slim people. The most worrying trend is an increase in what's called 'morbid obesity', when the BMI exceeds 40. Overall 6.3 per cent of US women – that's one in 16 – are morbidly obese.

The fact is, obesity is a killer. Every year it causes 300,000 deaths, costing the US healthcare system around $100 billion. Obesity leads to diabetes, strokes and heart attacks. It shortens careers, relationships and lives. And it's not only adults who suffer; there have been dramatic rises in child obesity rates as well.

Some see it as payback time for the nation that brought us car-choked streets and the double cheeseburger with large fries. The cause of the epidemic is most likely to be a combination of prosperity, cheap fat-laden food, car ownership and a world designed to minimise exercise. Try walking to the shops in the majority of towns and cities in the United States. Most stores are located in malls several miles away, and the only way to get there is by car. Parking lots are designed for customers to drive between shops rather than park in one place and walk. It's all made easier because petrol is cheap. Once inside the supermarket, the shopper is greeted by aisles of processed foods, snacks and sugary treats. 'Bad' food is both cheap and plentiful.

But just as getting fat is big business, so is getting slim. Consumers are bombarded with slimming products, exercise machines and weight-loss gimmicks on 24-hour shopping channels. It's easy to criticise the people who buy them. After all, if they weren't wasting their lives watching cable they could be taking some exercise. Like the right to carry arms, fat is part of American culture. But take a look in the mirror before you cast the first stone.

sedentary lifestyle we've become so attached to. In order to be able to do this, it's important to understand what's making us fat in the first place.

Weight gain doesn't happen overnight; it creeps up on us gradually over a long period of time. As we said above, it is the product of the amount of energy that we take in from food and drink exceeding the amount of energy our bodies use up. Ideally, we should take in the same amount as we expend. This is what doctors call a perfect energy balance. Weight gain comes about as a result of positive energy balance, when excess energy gets stored in the body for future use. The problem is we don't often get to use it.

Energy is measured in calories or their metric equivalent, joules. To give you some idea, women are advised to eat up to 2,500 calories per day, and men up to 3,000, but obviously these requirements depend on lifestyle. In food terms, this calorie quota is equivalent to just a single fast-food meal of double cheeseburger, large fries, cola and ice cream. A few of those a week and it's not hard to see where your body is heading.

The good news is that the human body needs an enormous amount of energy just to remain slumped in front of the TV or over a computer keyboard. If the most we do is stagger out of bed, visit the bathroom and then sit on a chair watching telly all day, we will use up almost half our recommended daily calorie intake. This energy use is referred to as the body's resting metabolic rate and the energy goes into maintaining normal functions such as breathing, cell division and circulation. Encouragingly, the fatter an individual is, the greater their resting metabolic rate, so an obese person uses up more energy doing nothing than a slim person. Another bodily function that eats up energy is a process called thermogenesis. This keeps us warm and results partly from the digestion of our food. It's also linked to a source of the body's energy stores, brown fat, which generates heat. Unfortunately, on a normal diet, keeping young and beautiful requires exercise. Not only does physical activity use up calories but also a body with developed muscles consumes even more energy at rest.

Just small changes in energy balance can tip us over the edge into the long slow slide towards obesity. But why can some people feast on fast food and not put on weight while the rest of us just sneak a casual glance at a gateau and gain 2 kilos? Why do some people have a different energy balance from others? There are a number of factors, some of which science might be able to do something about.

EATING ALL THE PIES

We tend to believe that we eat when we're hungry, but in reality that doesn't happen very often. Mostly we eat because it's about the right time for a meal. Another misapprehension is that we stop eating when we're full. It is now apparent that the process of appetite control is a lot more complicated than scientists first thought. It involves the conscious and subconscious parts of our nervous system, hormones and proteins. And it's wide open to abuse.

The central control unit for hunger and appetite is the hypothalamus, a cluster of small nodules tucked into the centre of the brain. Before we talk about what *does* happen, it's worth understanding what *should* happen. The hypothalamus is constantly monitoring the state of the body. It detects the amounts of glucose and minerals circulating in the bloodstream and receives signals from the digestive system and fat stores. If supplies are looking a little low, it sends a signal to conscious parts of the brain registering hunger. The reverse happens after a meal. The body signals that it's had enough food, and the hypothalamus lets the conscious brain know. And eating stops.

If it were that simple we'd all be the right weight and wouldn't keep buying bars of chocolate. Unfortunately for our beauty regime, there are a number of other factors at work. One of the most interesting is leptin. Dubbed the fat hormone, leptin is continuously produced in the body's fat cells and is secreted into the bloodstream. The hypothalamus monitors leptin levels to keep track of the body's fat reserves. But as we get fatter, research has shown that the hypothalamus becomes insensitive to the hormone's effects. Work done by Luciano Rossetti at the Albert Einstein College of Medicine in New York suggests it only takes a few high-fat meals to completely mess up the

Diet pills

Wouldn't it be wonderful if the diet pills advertised in Sunday supplements actually worked? If all we had to do to keep trim was suck a lozenge containing a few homeopathic ingredients, everyone would be a whole lot thinner. No more failed diets or lapsed exercise regimes, just a pill a day and we could all binge happy in the knowledge we'd still be able to fit into our clothes tomorrow.

Sadly, most diet pills only work for the shareholders of the companies that produce them. There are a few genuine weight loss pills on the market but only doctors can prescribe them. The two currently licensed in the UK are Xenical and Reductil. Xenical works by interfering with the functioning of the enzymes that digest fat in the gut. As a result less fat is absorbed into the body. Reductil acts on chemicals in the hypothalamus, inducing a full feeling after eating less food than usual. Neither drug is without side-effects and a recent report by the Royal College of Physicians warns that they should be given only to patients who are prepared to diet and take exercise. That mantra again. Not listening, not listening, not listening ...

We need a lazy-lifestyle pill – a safe drug that will banish weight gain for ever. Many drugs companies are investing in chemicals that interfere with the energy processes of individual cells. All the body's energy is generated at a cellular level – there's no central powerhouse. Cells take in molecules of sugar, fat and protein from the bloodstream and use them to generate a chemical called ATP. This is the molecule that powers everything in the body. If more energy is produced than we need, it is diverted into fat production and gets stored in fat cells. If scientists could find a drug that allowed cells to break down food but not produce ATP, they'd be on to a winner. Then by popping a pill we could eat as much as we wanted without putting on any weight at all.

It turns out that leptin (the hormone that the hypothalamus uses to monitor the amount of fat kicking around the body) might help us achieve that goal, because it also plays a role in burning off fat. It does this by activating an enzyme called AMP kinase that enables the body to use its stored energy. The discovery of this mechanism was made by scientists working for the UK Medical Research Council and Harvard Medical School in the US. The researchers also found that

AMP kinase decreases the amount of sugar in the blood. The enzyme is activated naturally during exercise within muscle cells, which raises the possibility of kidding the body into thinking it's doing some physical work. Imagine, you could be watching TV while your body thinks it's playing squash. The researchers believe that new treatments can be developed to artificially stimulate AMP kinase to burn excess fat. Because the enzyme helps control blood sugar, such research might also lead to new diabetes therapies.

So where does that leave our diet pill? Considerable resources are being put into developing drugs to combat obesity. Although the ethics of developing a binge pill for the motivationally challenged are slightly dubious, it would be an extremely lucrative market.

system. Rossetti conducted a series of experiments on rats and found that after 72 hours of eating a high-fat diet, the animals' ability to respond to leptin was completely lost.

Then there's sugar. Chocolate, ice cream and cake are undoubtedly some of humanity's greatest achievements, right up there with aircraft, antibiotics and the toilet seat that plays a medley from *Evita*. 'Sugar junkies' is a widely used phrase that may contain a grain of truth. Psychologists at Princeton University, led by Bart Hoebel, allowed rats to binge on sugar and then removed it from the rats' diet altogether. The rodents were sent into fits of anxiety, experiencing chattering teeth and the shakes. Not unlike the cold turkey symptoms induced by drug withdrawal. The study, published in June 2002 in the journal *Obesity Research*, seems to demonstrate that the rats were becoming addicted to sugar. Dr Hoebel suggests that this is because sugar triggers production of opioids, the brain's natural pleasure chemicals.

Research like this and studies of leptin have led some scientists to conclude that high-fat, high-sugar diets are addictive. This means that once we start, we'll need more than self-control to keep ourselves from raiding the fridge. What it doesn't explain is why some of us eat more food than we need to in the first place. Which is why the search is on for the genes for obesity.

The human genome

Every cell in the body – from the neurons in our brain urging us to have another chocolate to the fat cells accumulating around our waist when we do – contains the complete set of instructions for how to build a human being. This is the genome, the entire list of genes required to assemble not just anyone, only you. It takes more than 30,000 genes to construct a human.

The international effort to read this code, the Human Genome Project, began in 1990. Publicly funded, it has been described as 'the biological equivalent of putting a man on the moon'. At a press conference in June 2000, scientists announced that the entire genome had been sequenced. Actually it was the draft version; two and a half years later they had really finished it.

The genetic sequence is the order in our DNA of the four nitrogen bases (see page 10): G, A, T and C. There are around 3 billion pairs of them in our 46 chromosomes, so you can start to appreciate the scale of the undertaking. Yet almost 97 per cent appear to be meaningless, referred to by scientists as 'junk DNA'. It might turn out this isn't entirely the case and this DNA contains some secret message. In the meantime researchers are concentrating their efforts on the remaining 3 per cent, which contains our genes.

Despite our physical differences, every human on the planet is very nearly genetically identical. Only a few genes make us unique, and differences can often be down to a single letter in the code. Scientists refer to these as 'single nucleotide polymorphisms', or SNPs (pronounced 'snips'). Different letters found at a SNP can determine the difference between your eye colour and that of your partner's. It's these same single letter differences, or mutations, that lead to many genetic diseases.

By comparing the genomes of two different people, scientists can examine where SNPs crop up and, as a result, work out what they do. One individual might have a disease believed to be genetically inherited. By comparing their genome with that of a 'normal' person, scientists can understand which genes are implicated in the disease. The more genomes compared, the more researchers find out. This isn't just true of humans.

Because our genome is so similar to that of other mammals, it can be examined against theirs. A great deal of research is done in mice, for instance. Because they know the mouse genome, scientists can carry out genetic experiments on mice and apply the results to humans.

Having a list of all the genes in the human genome is undoubtedly useful. But it's not *that* useful. Genes are the code for particular proteins. These proteins form the basic building blocks of each cell and ultimately the entire body. A different set of proteins is found in every different type of cell in the body. So in some cells a gene might generate a protein, in others it won't. Producing a map of which genes produce which proteins is the next hurdle to overcome. The Human Proteome Project will enable the human genome to be put to use.

BORN TO BE MILD

People give any number of excuses for their weight – from 'I've got large bones' to 'I've got a hormone imbalance'. Either of which could be true but more often isn't and merely serves to excuse an extra portion of Häagen Daas. Let's face it, we all want to hear that weight gain is a failure of our biology not of our self-control.

Whether obesity is pre-determined by our genes or a consequence of our lifestyle is a hot topic of discussion among researchers. Lifestyle certainly plays an important part. If you eat a lot and take no exercise, the outcome is pretty obvious. But like many human traits, obesity is most likely to be a combination of both nature and nurture.

Over the last century, as scientists have started to get to grips with genetics, it has become increasingly obvious that the human genome is more complicated than they originally hoped. Although it's often likened to a blueprint to build a human, it's not laid out as a simple diagram. Like the animal it represents, the genome is a product of evolution.

There's no one single gene for anything much. There's not, for instance, a gene for happiness or a gene for mild irritability. It certainly seems that some babies are born happy and others grumpy so there might be some genetic factors at work, but it would be a whole load of genes acting together, rather than just one. *Some* things are determined by just a single gene, but scientists have found that usually our body chemistry is controlled by a combination of several or many. This is certainly true of obesity. Latest studies suggest that at least 200 genes are implicated in weight gain.

MONSTER MICE

Fat kids tend to have plump parents. A fact easily explained by lifestyle factors: if the parents overeat and don't get out much, then the chances are the children will follow the same behaviour patterns. But what actually happens is more interesting. Adopted children tend to have similar weight patterns to their *natural* rather than adoptive parents. Evidence certainly for a genetic basis for obesity, but which genes are they?

The first inkling that there might be a fat gene came more than 50 years ago, three years before the discovery of the structure of DNA. In 1950, scientists at the Jackson Laboratory in Bar Harbour, Maine, identified a genetically distinct type of mouse. They called it the obese, or *ob*, mouse, because it was double the weight of your regular mouse. Scientists at the Rockefeller University in New York identified the gene that made the mice so fat, the *ob* gene, in 1994. The same team, led by Professor Jeffrey Friedman, discovered what the gene did a year later. The *ob* gene codes for the protein leptin. It was the first time researchers had come across this protein. They named it after the Greek word for thin.

As we've discussed, leptin is the hormone that regulates body weight by signalling to the brain the amount of fat being stored. Mice without the *ob* gene weren't producing any leptin so weren't receiving a signal in their brains to stop eating. As soon as a mouse with the defective gene was given an injection of leptin the pounds fell off. The biology of mice is very similar to the biology of humans, but there has been much debate over whether what applies to mice applies to men.

Immediately following the publication of the *ob* gene research there were high hopes that leptin could be the miracle cure for obesity. In 1997, geneticists at Addenbrooke's Hospital in Cambridge carried out studies on two severely obese children. They found that both carried defective copies of the *ob* gene so were unable to produce leptin. The absence of the hormone was making them fat.

Although the *ob* gene can explain a few cases of obesity, other genetic factors are also important. In 2000, researchers at Harvard identified two molecules, GATA-2 and GATA-3, that appear to regulate the formation of fat cells. A failure in their structure caused by a genetic defect could cause the process to get out of control. More recently, in March 2003, another leptin-related gene was identified. This one, the MC4R gene, codes for a protein receptor on cells in the hypothalamus. Without it, leptin is ineffective – the brain doesn't receive a 'stop eating' signal and keeps going. Researchers led by Stephen O'Rahilly in Cambridge found that children with a MC4R gene mutation chose to eat three times as much breakfast as those with a normal MC4R gene.

NEW GENES

These are just some examples of the genetic factors involved in obesity. Of course, not all of us struggle to fit into our jeans because of our genes, but genetics certainly play their part. Over the last 50 years obesity has become much more of a problem; however, during this time the human gene pool has remained more or less the same. This suggests that all the genetic factors involved in obesity have been around for much of our evolutionary past. Then, what with all that hunting and gathering, they didn't really matter. It's only nowadays, with our increasingly sedentary lifestyle, that these genetic factors are starting to become apparent. But of course we don't want to change our lifestyle, so can we change our genes?

Gene therapy is a remarkable technology with tremendous potential. It gives us the opportunity to alter our own DNA, re-engineering the genetic blueprint in each of our cells. Changing a gene that helps make us fat, altering a gene that makes us bald or tinkering with a gene for nostril hair – all possible with the power of gene therapy. Not born perfect? Gene therapy can help you achieve that goal.

While genetic perfection might be your priority, for doctors, gene therapy offers the potential to cure genetic diseases. In most cases these are some of the more unpleasant ailments to affect humanity. We should say at the start, though, that gene therapy is still very much an experimental treatment. It's inherently dangerous and when it goes wrong the results can be tragic. You have been warned.

Diet another day

From Bond villains to latex-clad superheroes, gene therapy is where it's at when it comes to the appliance of science on the big screen. In the past, genetic mutations of the dangerous-experiment-gone-wrong variety usually involved radiation. These days the techniques employed by mad scientists are much more subtle. In the first of the latest *Spiderman* films our hero acquires his new abilities from the bite of a genetically engineered spider. Crucial parts of the arachnid's DNA are incorporated into Peter Parker's teenage body. He's not alone; the villain is genetically transformed too, but into a green monster, in a sort of green-monster-transforming machine.

Then there are those genetically tweaked to perfection in laboratories by scientists playing God. When will they ever learn? It's a theme exploited by TV companies in series like *Dark Angel*. Designed to appeal to a teenage audience, this programme features a cast of out-of-work models from the shallow end of the talent pool being hunted down by secret agents. In this case our heroes have been genetically bred as part of some never wholly explained government experiment. They all have barcodes on the back of their necks, so are in trouble if they ever get cornered in a supermarket.

Tinkering with an individual's DNA has also proved irresistible to the writers of the James Bond film *Die Another Day*. 007 stumbles upon a secret medical facility on a Cuban island where people seek genetic enhancement. The treatment is portrayed as long and painful, designed for the super-rich as an alternative to plastic surgery, although it was so effective that gene therapy could alter the villain's appearance from North Korean to north London. Certainly this scenario is closer to reality than most of the other fictional applications of genetics.

Any gene therapy is certainly fraught with hazards. Whereas plastic surgery alters only exterior appearance, gene therapy has the potential to reconfigure the entire body. Whether it could drastically rearrange facial features is a moot point, but there are certainly ways of using the technology for superficial ends. Although Cuba is reputed to have one of the best healthcare systems in the world, it's unlikely that the secret hospital really exists. Shame really – for our purposes it could prove very useful.

The idea of altering an individual's DNA to combat disease was first conceived back in the 1970s. Initially, a few partially successful experiments were tried on mice. But at the back of many scientists' minds was the thought that gene therapy could be used to cure genetic diseases in humans. At present there are thought to be at least 2,000 diseases caused by genetic defects. The abnormal pieces of DNA responsible for these conditions are known as gene mutations. Some are inherited, others are caused by external factors known as mutagens. Thousands of dangerous chemicals are mutagenic, which also makes them carcinogenic – causing cancer. Other genetic illnesses include cystic fibrosis and Huntington's disease. Some conditions are caused by mutations in just a single gene, others by mutations in a combination of genes. A mutation can be a missing bit of DNA or an extra bit of DNA, either of which codes for imperfect cells.

Before doctors can attempt to put any genetic defects right, they have to know which genes are implicated in the disease. Identifying these specific pieces of DNA is a process known as gene cloning. In this case the cloning isn't so much a copying operation as a DNA identity parade – sorting through the genome to identify specific genes. The power of gene therapy is also its fundamental weakness. Interfering with DNA can certainly, in theory, cure. If it goes wrong it can kill. Not surprisingly, many ethics committees have been reluctant to approve human trials.

In the case of some diseases, the patient is likely to die anyway. In 1990 approval was given for the first human trial at the National Institutes of Health in Bethesda, Maryland, in the USA. It involved treating a three-year-old girl with severe combined immune deficiency (SCID). This is a condition caused by a defect in a single gene that renders the sufferer's immune system completely useless. The gene in question is known as ADA and is essential for the formation in the bone marrow of T-cells, the body's white blood cells that fight infection. Unless they can be treated, children suffering from SCID do not usually live long. What life they have is spent isolated from infection and thus from the world, which is why the media often refer to them as 'bubble babies'.

At the time of the first gene therapy trial there were two other treatments available for SCID. One was a total bone marrow transplant from a matched donor – a painful and unreliable process. The other was a technique called protein therapy. Rather than change the gene, doctors injected the missing protein, allowing the T-cells to function. The problem with protein therapy is that it usually requires a lifetime of regular bone marrow injections. Despite these alternatives, scientists French Anderson and Michael Blaese were given the go-ahead to treat the three-year-old girl. More on that treatment in a moment.

VIRUSES THAT CURE

While the idea of replacing a faulty piece of DNA is simple enough in theory, in practice fiddling around with a tightly wound molecule inside a cell is, to say the least, tricky. In textbooks the genetic code is written out as a series of letters, representing the base pairs. Sadly, evolution hasn't generated similar markers on the real thing. Scientists can't just look down an electron microscope, identify the mutated DNA, snip it out and glue the genome back together. But there are things that can – viruses. Or, more specifically, retroviruses.

Viruses replicate by hijacking the machinery of their host cells. They use the infected cells for replication and protein synthesis. Retroviruses have their own particular way of doing this. The genome of a retrovirus is not described

Life unzipped

It's all very well harping on about DNA being the blueprint for life, but to be any use cells have to be able to read the instructions and build from them. DNA is the code for proteins, or, more accurately, the code for the building blocks of proteins, amino acids. Each set of three base pairs codes for one of 20 amino acids. Whatever it says on your bottle of moisturiser or breakfast cereal, all of them are 'essential'.

In a cell the DNA is contained in the nucleus and the proteins assembled in a part of the cell called the ribosome. Which suggests that something has to ferry information between them. This messenger is a molecule called ribonucleic acid, or RNA. It's usually called messenger or mRNA, because that's what it does. Messenger RNA is very much like DNA in that it's a long thread of bases – only it's a single rather than double strand. So how does it work?

The first part of the process is called transcription. An enzyme called RNA polymerase unzips the strand of DNA and begins to build a string of mRNA. Each base on the DNA strand has a corresponding mRNA base. For instance, base A (adenine) transcribes to base U (uracil) in the mRNA. Once the process is complete and a 'stop' code has been reached, the DNA zips back together, remaining totally unchanged. The mRNA can now take its information away.

The second part of the process is termed translation. This takes place in the ribosome, where the mRNA is translated into amino acids. Sequences of particular bases in the mRNA code for one or other of the 20 possible amino acids. The cell then goes about assembling these into proteins. Although every cell has every gene, different types of cell produce different proteins. It's the control of the process of protein synthesis that makes the difference between one cell and another. Special regulator proteins ensure that amino acids are only produced when they are needed.

in DNA, but rather by a fragment of RNA – the single strand molecule used by cells for protein synthesis. When a retrovirus infects a cell, its aim is to reproduce. The host cell is geared up for dealing with DNA not RNA, so the virus has to get its RNA converted. It does this by reversing the process of transcription, using an enzyme called reverse transcriptase. The RNA from the retrovirus is translated into DNA and incorporated in the genome of the host. The HIV virus is just one of many viruses that works this way.

For the purposes of gene therapy, retroviruses are the perfect way of delivering new genes into cells, as the RNA can be engineered to contain the corrected gene. Once injected it will do all the work of incorporating that gene into cells. To treat the child suffering from SCID, doctors extracted some white blood cells, infected them with a retrovirus containing the functioning ADA gene and re-injected them back into her body. The treatment worked straight away, the new gene going into action to produce fully functioning T-cells. Although she continued to have protein therapy, scientists had demonstrated the potential of using retroviruses to re-engineer the human genome.

You might wonder why, if the treatment was first developed in 1990, more genetic diseases aren't being tackled in this way. Certainly there was tremendous optimism following the treatment's initial success. The trouble is, although gene therapy is fiendishly clever, it's not always precise. Once the retrovirus is introduced, things can go tragically wrong.

LETHAL EXPERIMENT

Since 1990 thousands of patients had been treated using gene therapy and scientists seemed to be making slow but steady progress. Then, in 1999, 18-year-old Jesse Gelsinger agreed to take part in a gene therapy experiment. Jesse suffered from a rare liver disorder but, despite having to take a daily cocktail of drugs, was able to lead a fairly normal life. On 12 September he arrived at the University of Pennsylvania, where the trial was being held, and on the 13th he was injected with a genetically modified virus. Five days later Jesse was dead. He had suffered a massive immune reaction. As a result there was a worldwide reappraisal of the technology. Nevertheless, other trials that seemed to be producing encouraging results continued. Until there was another setback.

At the Necker Hospital in Paris doctors were using gene therapy to cure another type of severe combined immune deficiency disease, X-SKID. Eleven boys were being treated – until suddenly two developed leukaemia. In both cases the cancer emerged because of a fundamental problem with retroviruses: the scientists can't control where in the genome the new DNA is inserted. The corrective gene appears to have been incorporated near another gene called Lmo2. The Lmo2 gene helps regulate cell growth but if turned on at the wrong time this can go out of control, leading to cancer. This might only happen with the particular retrovirus used, but it's a considerable risk to take.

Until gene therapy is perfected, the current consensus is that it should only be used as a last resort. These tragedies will certainly lead to refinements in the techniques. Scientists are looking at ways of ensuring that the retroviruses only insert DNA into specific areas of the genome. They are also looking at ways of building in self-destruct mechanisms so they can easily kill the retroviruses if something goes wrong. Once a virus is injected into the body, it will replicate. Being able to stop it is probably a sensible precaution. That's not to say the scientists don't know what they're doing; gene therapy is only used to target specific tissues under carefully controlled conditions. It's just that it's far more complicated than they at first imagined.

BEAUTY THERAPY

So where does this leave our quest for genetic perfection and weight loss? How can we turn gene therapy into beauty therapy without undergoing radiotherapy? Despite the setbacks, gene therapy has the potential to cure genetic disorders such as cystic fibrosis, Alzheimer's or even cancer. But it also has the potential for genetic enhancement – re-engineering our imperfect bodies, the new plastic surgery for the super-rich.

It's already possible to be genetically tested for a whole range of different diseases. As scientists learn more and more about the function of each gene in our genome, the screening will get more and more sophisticated. If you find something you don't like – a mutation for the *ob* gene for example – gene therapy offers the chance to do something about it. Want blond hair? Shave off your highlighted mousy hair and get a new gene injected. As scientists decipher our genome and proteome, almost anything is possible. If not always ethical.

All the gene therapy treatments developed so far target very specific cells – invariably those in the immune system. This means that the original genetic defect can still be passed on to future generations. The solution is to genetically alter cells in the reproductive system – sperm and eggs. You could also, of course, re-engineer reproductive cells to produce a genetically 'perfect' child. This is known as germline gene therapy. Where gene therapy becomes eugenics.

ENGINEERING THE MASTER RACE

The word eugenics is probably associated in most people's minds with the atrocities committed by the Nazis before and during the Second World War. But the Germans were not alone in embracing such a deeply flawed science. Some Western countries still had pro-eugenics laws in place as late as the 1970s.

British biologist Francis Galton, the first cousin of Charles Darwin, came up with the term eugenics in 1883. His aim was to increase the sum of human happiness by improving inherited characteristics. This idea of racial improvement would encourage the 'well born', so that only the ablest and healthiest people would have children. Although Galton is often portrayed as a villain, he only ever advocated 'positive eugenics', encouraging the 'right' people to breed rather than preventing the 'wrong' ones from doing so. Which is ironic considering that the British royal family at the time was suffering from some of the worst genetically inherited diseases (such as haemophilia) after generations of inbreeding.

Scientists, philosophers and politicians embraced the philosophy of Galton and his advocates and by the 1900s eugenics was becoming increasingly popular in Britain, mainland Europe and the United States. The prevailing scientific wisdom at the time was that pretty well everything was inherited, from eye colour to mental ability. It seemed perfectly reasonable to many at the time that the Darwinian theory of 'survival of the fittest' meant that only the fit should be allowed to breed.

By 1911 six American states had laws allowing the forced sterilisation of the mentally ill – the law in Virginia was only repealed in the 1970s. By the 1920s, the Eugenics Records Office at Cold Spring Harbor Laboratory in New York had carried out a series of studies seemingly backing the new science. Research suggested that inherited traits included 'shiftlessness', which made people poor, and 'criminality'. Advocates of a similar policy in the United Kingdom included Winston Churchill, George Bernard Shaw and H. G. Wells, but remarkably no pro-eugenics law was ever adopted. The proposal that the government should have such intimate control over individuals' lives was successfully fought in Parliament on libertarian grounds. Not so in the land of the free. In the United States thousands of immigrants were barred from entry because of their genetic heritage (such as skin colour), a situation that continued for close on 40 years.

Few countries can escape an association with eugenics. Neither can scientists distance themselves from the practice. It was, after all, science that provided all the evidence politicians needed to adopt 'racial purity' programmes. But eugenics itself is long discredited and no democratic government would adopt a public policy of sterilisation. Instead, individuals make the choices. Technologies such as gene therapy and therapeutic cloning give us the opportunity to alter our own or our children's genes. Genetic testing, for instance, means we have a choice over whether to abort a foetus we know will be born severely mentally retarded. The quest for genetic perfection continues. In a recent interview to celebrate the discovery of the structure of DNA, Nobel Prize winner James Watson commented, 'If scientists don't play God, who else is going to?' Eugenics isn't dead. It's just got more complicated.

At the moment gene therapy is an extremely risky undertaking. Messing around with retroviruses in your home laboratory is probably ill advised, and injecting them certainly a foolhardy undertaking. Once the technology is refined though, gene therapy could put a lot of plastic surgeons, personal trainers and health gurus out of business. Whether it will lead to the creation of a genetically perfect super-race is debatable. But if you knew your child was going to suffer from an inherited disease, wouldn't you do everything in your power to fix it?

As for gene therapy as a weight or cellulite treatment programme, it's probably best to wait a few years. So if changing your body is out of the question, how about changing your food? The obvious option would, of course, be to follow the advice of nutritionists and eat a healthy, well-balanced diet. But that involves a degree of self-discipline we're not prepared to countenance. The alternative is to scientifically re-engineer foods so that they not only taste better but contain fewer calories. For instance, we might create a chocolate cake that looks and tastes fantastic, contains lots of healthy vitamins and minerals but doesn't make us fat. This means we need to stop worrying and learn to love genetically modified foods.

SEEDS OF PROGRESS

Genetically modified, or GM, foods promise a great deal. Higher yields and reduced workload for farmers, fewer herbicides and pesticides, and better products for the consumer. Of course, if the argument were that simple environmental groups wouldn't have invested so much effort in campaigning against them. In the United States around a third of maize (or corn, as it's known in America) is genetically modified as is around 80 per cent of the soya bean crop. That's tens of millions of acres planted each year. Most of these crops are what the industry terms 'first generation' – genetically enhanced to benefit the farmer with traits such as herbicide resistance and immunity to particular diseases.

One of the first commercially successful products was Monsanto's Roundup Ready series of seeds. Roundup is an effective, safe and relatively cheap herbicide that the company had been producing for some years.

Monsanto decided to engineer soya beans tolerant to the chemical. That way a farmer could spray the entire field, the weeds would be killed and the crops would remain untouched. In theory this should reduce the number of sprays necessary, thus minimising labour and costs to the environment. Crucially for Monsanto, the farmer now not only buys the herbicide from the company but the seeds as well.

Monsanto is one of several multinational chemical companies that in the last few decades have diversified into biotechnology. When scientists started taking the first tentative steps in plant genetic engineering, many businesses saw this as a lucrative source of future income and invested millions, if not billions, of dollars in research. But achieving commercially viable GM crops has proved to be much more difficult than many at first hoped and along the way a lot of the smaller biotech companies have gone out of business. Even executives at Monsanto were unsure at times whether the goal they were pursuing was worthwhile. Nowadays, at their headquarters in Saint Louis, they preach GM technology with an almost religious zeal.

The Monsanto laboratories sit in a wooded valley on the outskirts of Saint Louis – a group of enormous five-storey brick buildings surrounded by landscaped grounds. 'Beware Deer Crossing' signs dot the curving roadways. The appearance of the labs is somewhat at odds with the company's public image – it is portrayed by environmental groups as an 'arrogant' organisation bent on world domination. Inside, the proud history of the company is illustrated on a timeline on the wall. Monsanto started out around 100 years ago making caffeine for Coca-Cola. Not surprisingly, there's no mention of the fact that it once also produced Agent Orange – the defoliant used to devastating effect during the Vietnam War.

The labs at Monsanto are surrounded by rows of greenhouses, where the latest biotech crops are being put through their paces. What's being cultivated at the top of the building is the end product of a manufacturing process, rather as in a more traditional industry. On paper it's a relatively simple procedure: decide what you want to do, find the gene that you need, insert the gene into the cells and grow the resulting plant. So if you've got that scientific super-market reward card ready, let's go shopping.

HOW TO GROW A GM CROP

Here's what you need if you want to take on the likes of Monsanto and develop your own GM maize:

- ❀ Half a dozen healthy maize plants (NOT tinned sweetcorn)
- ❀ Any other animal or plant whose genome you think might be useful
- ❀ Assorted test tubes, petri dishes and a microscope
- ❀ A box of shotgun cartridges and pile of gold dust (honest)
- ❀ A sunbed or a sunlamp
- ❀ A small greenhouse
- ❀ A good lawyer

You'll probably also need some sort of government licence and police protection, but let's not make it too complicated. From start to finish it takes a well-resourced organisation many years to develop a new crop, so you'll also need plenty of spare time.

The easiest part of the whole process is deciding what you want to change. Perhaps, like Monsanto, you've developed a best-selling herbicide and want to create a resistant plant to go with it. Maybe you want to engineer a built-in insecticide that kills any plant pests that chose to munch on your maize. Or maybe you want to create a crop that's blue instead of yellow, purely for aesthetic purposes. Or because you're an Everton fan. The easiest bit – choice of colour – is rapidly followed by the hardest bit, identifying the gene to do the job.

As we've said, it's not easy to find single genes that confer specific traits. Just as around 200 genes are implicated in obesity, many different genes can control plant height, taste or colour. Identifying the right tiny section of DNA is time-consuming and laborious. Even once it's discovered there's no guarantee that someone else hasn't got there first and filed a patent (hence the lawyer).

HOW TO PATENT THE PERFECT BLONDE

The idea of a patent is to protect an invention. The meaning is pretty obvious when you apply it to the Nelson-Hollingham tomato twizzler (patent pending) but not so obvious when it's applied to living things. To be eligible for a patent your invention must be new, innovative and useful; it can't simply be something you've discovered. Patent laws vary between countries and, despite worldwide agreements, international patent law is riddled with ambiguities and loopholes. Broadly, what applies in one country should apply elsewhere. In practice it rarely does.

Although new genes are discovered all the time, they can only be patented if they can be linked to some kind of use. In the case of human genes this usually means a medical application such as a diagnosis or therapy. Life that occurs naturally cannot be patented – so we can't file patents on ourselves. However, it is possible to file a patent for a genetically engineered organism such as a new variety of corn or a perfect blonde.

There's a good reason for patents. It costs an awful lot of money to develop a new product, whether that's a corn plant or a tomato twizzler (you won't believe the number of tomatoes that were wizzled instead of twizzled during initial research). Patents enable the inventors to recoup that money before everyone else jumps on the bandwagon. A patent is a contract between an inventor and government, granting protection to the inventor in return for full disclosure of the technology. So once our twizzler is patented you'll be able to see the breakthrough innovations employed.

A GM food, for example, can take years of costly research and testing, hardly a worthwhile investment if everyone else can copy your work once it's finished. With a patent, the inventors can make the money back by either manufacturing the product themselves or licensing it to others. Only occasionally do governments override these rules – if they believe the invention is sufficiently important.

Opponents of gene patenting argue that genes are discoveries and it's often difficult to determine the particular function of a gene. As a result genes are being patented with only a vague understanding of what they might do. There's also an argument that rather than helping research, gene patents impede its progress. Once a particular company or organisation has patented a gene, others may stop their studies in that area. Equally, just because a patent is filed doesn't mean that a product related to it is under development.

It's not unusual for several different companies to be working on the same genes, racing against each other to file patents. Anyone undertaking commercial genetic research to develop a new treatment or product needs to have a good lawyer.

The traditional way of identifying the right gene is to work backwards from the protein that gene codes for. A good example of how this works in practice is the creation of the first types of successful genetically modified crops. *Bt*-cotton and *bt*-corn are engineered to express a protein that kills certain types of insects. *Bt* stands for *Bacillus thuringiensis*, a species of soil-dwelling bacteria that naturally produces a protein poisonous to the larval stages of many insects (such as most caterpillars). By identifying the gene that coded for the insecticide protein, researchers hoped to splice it into plants. Making them insect-proof.

There was an unseemly rush among university academics and half a dozen biotech companies to get to the gene first. Several claims were made and court battles fought. It turned out that there was a single gene for the *bt* protein, which made the science a little easier but the litigation more tortuous. In the end, several different products emerged, Monsanto even went as far as re-engineering the gene to make it more efficient, inserting new bases in the DNA strand and removing others. But having the gene is one thing, getting it from the genome of a bacteria to the genome of a crop plant is quite another.

WILD WEST BIOLOGY

The process of transferring genes from one organism to another owes less to the multibillion-dollar biotech industry of the Midwest and more to the old Wild West. The most dramatic method utilises something called a gene gun. This is a gun that fires genes. Looking rather like a cylindrical metal coffeepot, a gene gun is divided by horizontal shelves into a series of small chambers. Attached to the top is a smaller cylinder, which is where you load the shotgun cartridge. The gene to be inserted – what's called the transgenic DNA – is coated onto fine particles of gold. When the gene gun is fired, the explosive force flings the gold particles at the target plant tissue, where they penetrate the nuclei of each cell. When plant engineer John Sanford first came up with the idea for a gene gun, not surprisingly he was laughed at. For Monsanto among others, the gene gun is now the weapon of choice.

An alternative method of gene transfer works in a very similar way to the viruses used in gene therapy. *Agrobacterium* is a naturally occurring bacterium found in the soil. Like the retroviruses, it normally causes disease, but its ability to transfer DNA into plants has led to it being dubbed 'nature's genetic engineer'. By re-engineering its DNA, harmful genes can be removed and new ones added. Then it's used to infect the plant tissue, incorporating the novel DNA into the plant's genome.

Whichever method you choose, if you follow the correct procedure you should end up with a few genetically modified plant cells. These cells have been, to use the correct term, transformed. Of course, at this stage it's impossible to tell whether your transformation has been successful. One way around this problem is to use marker genes. These are usually linked in some way to the transgenes and help researchers to check whether all's going to plan. Marker genes have properties that are immediately easy to spot, expressing a colour change in transformed plant cells, antibiotic resistance or even fluorescence. It's not unusual for transformed plants to glow in the dark.

THE RIGHT BREEDING

So how do you make a plant from a few transformed cells in a petri dish? Fortunately, nature is on your side. All you need is a little culture. Although it's been said that playing music to plants helps them grow, in this case you'll need more than a selection of light classics. Under the right conditions with the correct combination of chemicals, individual plant cells can regenerate into

Seeds of doubt

Mention GM to anyone in the United States and they'll most likely think you're talking about General Motors. Genetically modified crops are widely accepted and consumed. That's not to say that Americans always know what they're eating, as all attempts to get labelling of GM foods have met with stiff resistance. You'd be hard pushed to have a meal in the US without consuming GM ingredients in some form or other, from the food itself to the oil it's cooked in. And as the biotechnology industry is anxious to point out, no one's had so much as a stomach upset as a result of eating a genetically modified food. So why are campaigners so opposed to it?

There are two strands to their argument, one environmental and one economic. The biotech companies claim that their crops can only benefit the environment. Insect resistance in particular should lead to a substantial reduction in insecticide use. Pro-GM farmers also say that herbicide tolerance has reduced the need for spraying.

Inevitably, caterpillars will develop resistance to the genetically enhanced *bt* plants. Herbicide-resistant plants will crossbreed with natural varieties making them resistant as well and over time there may be other unexpected side-effects. One environmental scientist who's raised concerns about the effects of GM crops is Chuck Benbrook. He suggests there's evidence that these plants are having an adverse effect on the soil, genetically altering important soil bacteria. At the moment the growth of GM crops is very much a real world experiment. Only time will tell if any of these fears are well founded.

The economic argument against GM crops is perhaps clearer. Agricultural biotechnology companies such as Monsanto don't just own the patent for a particular crop, they effectively own the seed. Any farmer wanting to plant a GM crop has to buy the seed and the chemical that goes with it – and sign a legal agreement prohibiting them from saving any of the seed. Some activists say agreements like this turn 'farmers into serfs'.

In the meantime the big biotech companies have gone on something of a charm offensive. Monsanto even has a 'pledge', which talks among other things about respecting the consumer. Of course, not all GM crops are developed by biotech multinationals. Many scientists see genetic engineering as a way of helping alleviate world hunger.

entire plants. With GM experiments, usually the more cells you've got, the better, Certainly you'll need quite a few plants to ensure that the transformation works.

In the first instance the transformed cells are cultured in dishes, so the cells divide into an undifferentiated mass called a callus. Callus tissue consists of the sorts of cells that grow when a plant needs to repair an injury. Once a sufficiently large ball of tissue has grown, it can be pulled apart and placed on a growth medium, where plant hormones trigger the cells to start developing into a complete plant. These tiny shoots are placed in heated rooms under artificial sunlight to develop into little plants or plantlets. Eventually these can be transferred to pots and grown in a greenhouse.

Assuming it's all gone to plan, you're now the proud owner of a genetically modified crop. But the process is by no means over. You should probably now check whether you've produced a low-calorie wonder crop or a man-eating Triffid. If it takes to following you around and hissing, then suspect the latter and destroy it immediately before it colonises the Earth. If, on the other hand, it looks perfectly normal, it's worth carrying out some simple tests. Introducing new genes can have unintended side-effects. While the new gene might keep insects at bay, it may also make your plant inedible. The same properties may stunt growth or interfere with some other part of the plant's biology. Then, of course, there are the environmental and ethical considerations.

THE NEXT GENERATION

Almost none of the genetically modified crops on the market at the moment provide any direct benefits to the consumers that eat them. Instead they're designed to make cultivation easier and more profitable. GM corn still looks like corn and tastes like corn. This is why advocates of biotechnology talk enthusiastically of the *next* generation of GM crops – ones aimed at us, the consumer. If you're ever to lose your love handles, these are just the sorts of developments you need to welcome.

Many food crops are already conventionally bred to have a certain proportion of nutrients, such as proteins and carbohydrates. Genetic technology makes it possible to engineer nutrition – to insert genes that express increased amounts of vitamins and proteins and reduced amounts of sugars without affecting the taste. Just imagine, you could eat exactly the same foods in the same quantity yet get improved nutrition and fewer calories.

In India a new type of rice is about to go on the market genetically engineered to produce higher levels of vitamin A. It's called Golden Rice because the vitamin A makes the inner kernel turn yellow. Invented by Dr Peter Beyer and Professor Ingo Potrykus and developed by an international team of scientists, it's been designed to help combat an effect of malnutrition. Vitamin A deficiency is an enormous problem in poorer parts of the world, even leading to blindness. Golden rice is not designed to make anyone much money; it's been created to solve a specific problem. But rest assured, GM foods aimed at affluent, overweight Westerners won't be far off.

In the meantime, the agricultural biotechnology industry is turning its attention to drugs. In small isolated plots across the United States, crops are being grown to produce pharmaceuticals – in one instance corn engineered to produce a protein to help treat cystic fibrosis. It turns out to be much cheaper and more efficient to grow drugs in a field than to mix them in a test tube in a lab. Although profitable, 'pharming' drugs is extremely controversial. Already entire harvests from several farms have had to be destroyed because food crops have become contaminated with pharmed crops.

THE NELSON-HOLLINGHAM PRIVATE CLINIC

So how can all this genetic technology work for the average man or woman in the street who has, shall we say, 'issues' with the way their body looks at the moment? Perhaps, for example, they're overweight, have little self-control, eat too much and drink to excess. Let's, for the sake of argument, assume that they're also going bald, have a yellow tinge to their skin (suggesting the advanced stages of jaundice) and only have four fingers on each hand. Let's call this average person Homer.

Now any properly trained doctor would at this point prescribe a sensible balanced diet, encourage exercise and possibly suggest counselling. But let's take a quick fantasy trip to the other end of the ethical scale and, before you can say 'malpractice suit', sign Homer up for some genetic alteration. Remember to keep that lawyer's number handy.

Probably the most lucrative place to start is at the top with his practically bare head. Baldness is very often an inherited characteristic – you can be genetically predisposed to losing your hair at an early age. But because there are genetic factors at work, a little gene therapy could well prove beneficial.

Now for the rest of the body. Homer is a living, breathing product of what happens when you embrace the American dream, smother it in ketchup and serve it up with large fries. Although Homer might be predisposed to obesity, unless he's got a serious genetic disorder, gene therapy is unlikely to prove particularly effective. He needs to consume fewer calories, but in such a way that he doesn't notice. A future of delicious, nutritious GM food beckons. Beer made with genetically engineered yeast and barley – guaranteed to provide 90 per cent of your recommended vitamin intake but with fewer calories and a kick like a mule (which is a good analogy because a mule is one of humanity's earlier genetic experiments). Then there's the chocolate you can eat between meals without spoiling your appetite. As well as with meals, after meals and in the bath too, if you prefer.

Frankly, we're not sure what Homer can do about the yellow skin and four fingers on each hand. With any luck there will be some unexpected side-effects of the inadequately tested GM foods and he'll grow an extra limb and turn purple. That's the beauty of playing God with limited knowledge – you never quite know what's going to happen.

THE COST OF PERFECTION

Now that the sequencing of the human genome is complete, there's tremendous potential for new uses of genetic technology. All scientists have to do is find out what each gene does. And that could take some time. But there's certainly no fundamental reason why science shouldn't help you remove those love handles for good. In the meantime, GM foods are already available and future products promise great nutritional benefits.

OK, so gene therapy is a little hit and miss at the moment and there are a few environmental question marks hanging over GM crops, but who said progress wasn't without risk? Incidentally, we should remind you that the authors accept no responsibility whatsoever. For anything.

People with a genetic mutation predisposing them to obesity may find that gene therapy becomes a viable treatment in future. For others it's probably a little too radical, unless scientists discover a gene for self-control. Whether consumers choose to embrace the next generation of GM foods rather depends on who they choose to believe: the biotech industry or the environmental industry.

If scientific solutions are making you a little uneasy, there is, as they say, a third way. Eat less food, join the gym and get yourself a personal trainer. You never know, they may turn out to be your perfect partner. Alternatively, if you really can't be bothered, drive to the shops and buy yourself some more chocolate.

Because you're worth it.

CHAPTER FIVE

How to Turn Back Time

There is a game anyone can play within the privacy of their own thoughts. It's called If Only. No dice are needed, no minimum number of players, and the rules are simple. All you have to do is mentally catalogue your life and recall all those events that would benefit from an 'If only …' attachment. If only I hadn't failed that exam … If only I hadn't drunk that tenth pint … If only I hadn't slept with my best friend … If only I hadn't hit that policeman … Or, if you're really unlucky, all of the above.

Just imagine the giddy consequences of the If Only game. You could be in a different job or a different relationship; you could be wearing better clothes – or hardly any clothes at all if only you had followed your heart and become a sailing instructor in a hot climate instead of an accountant in a cold one. The possibilities are endless, but, as anyone who has ever played If Only will testify, it is also a dangerous game. Sure, it's never too late to change many things in life, but some incidents are beyond repair. To continue playing If Only when there is nothing you can do about it can lead to a lifelong dependency on Prozac.

It is unlikely, given the possibility of permanent depression, that the If Only game will ever appear in all good toyshops. Yet what if there was a way to turn back time and right those wrongs in life? Who was to know, for instance, that losing those love handles would ruin your chances of being a plus size model or a professional opera singer? What if we could all genuinely have a second chance? In fact, wouldn't it be great if science could turn the If Only game into reality. If only ... I could build a time machine.

Before anyone starts huffing and puffing that this is supposed to be a science book, remember this: there is nothing in the laws of physics to prevent someone travelling through time. But before we rush out and start building our own time machines, let us consider another travel-sized version of the If Only game first. Think of it as the all-region, multi-format version. Here's why: some scientists believe we may already be living our own, alternative If Only lifestyle. The good news is that the world where you married your childhood sweetheart may already exist. The bad news is, you're not in it.

PARALLEL UNIVERSES

The idea of a parallel universe is a beguiling one. Life is filled with choices and the decisions we make along the way affect the direction our lives take. Imagine, then, that there are many Earths, exactly the same as ours, where every possibility is taking place – each one in parallel to all the others and at the same time. In one universe you get married, in another you act on those pre-wedding-night nerves and scarper. There could be one universe where you learn to sail for fun and a separate universe where you win the America's Cup. Anything is possible.

These parallel universes, known collectively as a multiverse, arise from a branch of physics that is pretty whacky in its own right yet has so far been found to be correct. If you have read the Chapter 3, you will immediately know what we are talking about. Yep. Quantum theory. If you haven't read that chapter then stop being such a freestyle individual and go back and read it now. Please.

To recap, quantum theory is the science of the very small, where weird things happen that bear no relation to what you'd expect in our larger macroscopic world. Quantum theory introduces probabilities and a fuzzy indeterminate state called superposition. If you have a particle that could have two states of spin for instance – up or down – quantum theory says that until you start measuring it, that particle's spin can be up, down or a superposition of up and down. In this fuzzy state of superposition the particle has, to all intents and purposes, its own set of parallel universes.

According to quantum theory, these states of superposition are in existence only while the system is not being observed and measured. As soon as someone tries to measure a state then all these undetermined states collapse in your universe so that only one outcome remains. At least that's what some scientists believe is happening.

In 1957 a physics graduate student at Princeton University decided to propose an alternative point of view for his PhD. A pretty risky subject for a thesis when you think about it, but luckily Hugh Everett's supervisor was John Wheeler – the physicist who later went on to devise the term black hole and wormhole. Wheeler was convinced that his student's work should be taken seriously and even published a paper alongside Everett's to add impact to his student's argument.

Everett decided to take Schrödinger's wave equation (the one that describes a particle as a wave or as a sum of multiple overlapping waves) at face value. He imagined a situation in which all the possibilities or states are real. Instead of collapsing on measurement, they continue existing in their own universe. In a state of superposition the particle has, to all intents and purposes, its own set of parallel universes. When you make a measurement, the particular outcome or result – spin up, say – becomes a part of your reality. The other possible outcomes, instead of disappearing, all have their own reality. In these separate realities the outcome of your measurement – spin up – is now one of the alternative possibilities instead. Everett called this the 'relative state metatheory' although luckily, thanks to Bryce de Witt, this became known as the much catchier 'many worlds' theory. It is an interpretation of quantum mechanics in which all realities exist in a multiverse rather than one reality existing in one universe.

Even though these ideas apply within a quantum framework – which means that the parallel universes are distinct only on a microscopically small scale – it didn't take much imagination on the part of science-fiction writers to ignore this fact and extrapolate the idea further. Since then, numerous

Parallel universes sci-fi style

First of all, special mention must go to three novels: Lewis Carroll's *Alice's Adventures in Wonderland* (1865) and its sequel *Through the Looking Glass* (1871); and C. S. Lewis's *The Lion the Witch and the Wardrobe* (1950). Though neither is science fiction, it must be conceded that when Alice stepped through a mirror or down a rabbit hole and Lucy walked through a wardrobe into the land of Narnia, where else did they go but a parallel universe? Not a quantum parallel universe admittedly – unless Alice's shrinking potion reduced her far more than we realised.

These days, television tends to produce the best versions of many worlds theory – no doubt because computer technology can deliver the goods with the necessary special effects. Parallel universes have been used in *Stargate SG-1*, through the introduction of a 'quantum mirror' in the episodes 'There But for the Grace of God' (season one) and 'Point of View' (season three). The quantum mirror allows the intrepid SG-1 team to see and visit alternative realities. In one reality, for instance, planet Earth is about to be taken over by an alien race called the Goa'ulds (don't worry about how it's pronounced – everyone seems to pronounce it differently in the series).

In season nine of *The X Files* an episode called 4D uses parallel worlds so successfully that you don't even miss Fox Mulder. The story begins with FBI agent Monica Reyes being murdered in a stairwell by a male psychotic. Agent John Doggett is then critically wounded by the same man with a gun. In the next scene, however, both Reyes and Doggett are alive and discussing the merits of Polish sausage. The telephone rings and Reyes is informed that Doggett has been shot. When she looks up, understandably confused, Doggett has disappeared. Worse still, the man who only we know is the killer identifies Reyes as the shooter, and Doggett, who is seriously injured in hospital, cannot understand why Reyes is alive after having seen her murdered. And neither, quite frankly, can we.

Luckily Reyes comes to the only possible conclusion: the other agent Reyes and Doggett must belong to a parallel universe and the killer knows how to travel between these worlds. The story is finally resolved when the killer is caught and shot dead (FBI agents in *The X Files* rarely wound). Reyes believes two versions of one person cannot exist in the same parallel world, so she pulls the plug on the paralysed Doggett after concluding that he is from the alternative reality. Her theory is that the other, healthy, Doggett will return. It's a risk but she does it and – phew – everything returns to normal.

There are also several episodes exploring this theme in *Star Trek: The Next Generation*, *Voyager* and the original *Star Trek* series. In 'Mirror, Mirror' (an episode from the 1960s version), Kirk, McCoy, Scotty and Uhura are sent into a parallel universe by bad weather and a transporter malfunction. Here, the Federation Empire resembles a Klingon organisation and previously trustworthy crewmembers have become unscrupulous alternative versions of themselves. Teleportation and many worlds – two fantasy science subjects for the price of one.

novelists have used the concept to create an alternative reality based on one key event in history. Some of the better results include Philip K. Dick's award-winning book *The Man in the High Castle* (first published in 1962), in which the Second World War has left America a divided nation occupied by the Japanese in the north and the Germans in the south. More recently, Robert Harris's book *Fatherland* is set in a world in which Germany's Third Reich is still in power and Hitler is about to celebrate his 75th birthday. In a multiverse, both these scenarios would exist at the same time as our reality. There was a television series, *Sliders,* based on the concept of a portal to a whole range of parallel universes. In each episode the characters found themselves in an alternative reality where different versions of their life existed. The show has now been cancelled, but we assume that in a parallel universe new episodes of *Sliders* are still being shown and it is now into season 17.

For decades Everett's parallel worlds theory attracted more attention from science-fiction writers than from other scientists – until, in a mind-teasing book, *The Fabric of Reality,* Oxford University physicist David Deutsch championed the idea of a multiverse.

Deutsch gives another explanation for the light and dark stripes that form an interference pattern when you shine a light through two parallel slits. The conventional reason is that light behaves as a wave and so overlapping waves sometimes cancel each other out and produce darkness. Deutsch suggests something far more 'out there'. He says the dark strips are produced when otherwise invisible 'shadow' light particles, or photons from a parallel universe, interfere with the light.

It may even account for the totally unexpected findings that result from sending light particles or photons through these closely spaced slits one at a time. With two slits open, a characteristic interference pattern of shadow and light results. Cover one slit and the photons will be travelling through the only open slit one at a time. As each photon will not have any other photons from another slit to interfere with, naturally you would not expect to see an interference pattern. And, when the individual photons pass through a single open slit, this is exactly what happens. You get a single bright spot on a detector screen, directly ahead of the open slit.

It's only when the number of photons passing through the slit is increased that things start to get weird. A pattern starts to emerge. The photons on the detector screen begin to line themselves up into bands of light and shadow and eventually, when enough photons have passed through the open slit, they form an interference pattern – exactly as if two slits had been open.

If Deutsch is correct and parallel universes are real, maybe we should take comfort. If you are unhappy about the direction your life has taken here, you can be assured that somewhere, in a different universe, your life is turning out exactly as you would have wanted. In one world you worked harder at school, became a doctor and were able to save your father's life when he had a heart attack after the Sunday roast. In another, you resisted temptation at an office party and avoided an acrimonious divorce and financial hardship.

Perhaps, too, we can eradicate one of the mind's greatest preoccupations – regret – because all the scenarios of the If Only game are being played out for real. OK, so it's the alternative you who will benefit from an alternative life in an alternative universe, but at least someone's life will be turning out well.

It is, no doubt, a reassuring thought – even though that same parallel universe may also be the one where David Beckham decided to become a hairdresser instead of a footballer, and Margaret Thatcher is still in power. On the bright side, as we shall see, the existence of parallel universes can also help solve some of the time travel paradoxes and rebuff those killjoys who say time travel can't be done.

THE TIME IN TIME TRAVEL

We are all time travellers. Everyone on Earth is travelling into the future. Unfortunately for the thrill seekers among us, this form of time travel is at the rather sedate rate of one second per second. Each of us is travelling into the future slowly and steadily, with no way of going back. Until someone builds a time machine.

So what is time and how do you describe a second, the smallest basic unit of time in everyday use? It is a simple question yet the right answer is not the obvious one. Think about how you define a second. If you're about to say one sixtieth of a minute, which is one sixtieth of an hour, which is one twenty-fourth of a day, which is the time it takes for the Earth to turn around on its axis once (or something like that) think again.

The answer, as agreed and defined in 1967 at the 13th General Conference of Weights and Measures, is '9,192,631,770 periods of the radiation corresponding to the transition between the two hyperfine levels of the ground state of the cesium-133 atom'. Not what we expected either. Time, as we now know it, is defined by the oscillations of an atom. It has not been defined by the Earth's rotation for more than 30 years.

A solar day remains the time it takes for the Earth to revolve once on its axis. This day is divided into 24 hours, each hour into 60 minutes and each minute into 60 seconds. But rather like a wind-up watch or the battery in a digital watch, our own planet cannot always be relied upon. The Earth's rotational speed can vary slightly and its axis of rotation is tilted. This leads to inconsistencies and this just won't do.

Mankind has spent thousands of years learning how to measure time accurately. Ancient Egyptian water clocks consisted of a water-filled urn with a hole near the base. It was used to time speakers in debates so that everyone spoke for the same length of time. Other methods have ranged from using stars, sand, shadows and marks on candles to the gears and springs of mechanical timepieces and the frequency oscillations of a quartz crystal. The atomic clock beats all of these for consistency and accuracy, which is why it is now the standard for measuring time.

The nature of time has also been redefined by science, and by one man in particular: Albert Einstein. Einstein wrote two major papers on relativity. The first, later called his Special Theory, appeared in 1905 and is all about time and space and what happens when particles move close to the speed of light. The second paper, published in 1916, is about the effects of gravity and is called the General Theory of Relativity. To understand more about time in our quest to build a time machine, we need to begin by looking at Einstein's first paper on relativity, the Special one.

EINSTEIN'S SPECIAL THEORY OF RELATIVITY

OK, before we get into the science, let's start with the most obvious question here. Why was Einstein's 1905 theory of relativity called 'special'? Particularly considering his next paper on relativity, 11 years later, was plain old 'general'. Was Einstein simply extra pleased with this one? Not exactly, because Einstein didn't even call his paper the Special Theory of Relativity. This just came to be the best way to describe it.

Probably because the original title of the published work is 'On the Electrodynamics of Moving Bodies'. Less snappy, yes, although it does give a better clue to what the paper is all about – basically things that move. To be more precise, the speed at which things appear to be moving from different points of view or, to use the scientific term, different frames of reference.

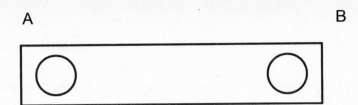

A   B

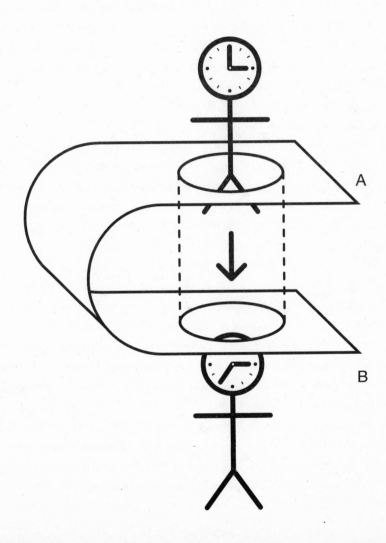

A

B

The 'special' in Special Relativity came about because Einstein's paper explained a special case of relativity in which any motion is uniform (it doesn't vary) and there is no acceleration. It also addressed the particular problem caused by light, because light didn't fit any of the previous predictions about what should be observed when light itself was in uniform motion.

Throwing a ball back and forth to someone across the park is a pleasant way to pass some time, so let's begin by using this simple activity to understand relativity. Unless you are a fast bowler for the England cricket team, the ball will probably travel through the air at a leisurely speed of, say, 2 miles an hour before your friend catches it. Now it's time to make this game more interesting.

Imagine there's a railway line beside the park. You take your ball on to a train at a station a couple of miles away. (We'll use miles rather than kilometres here because although science is metric, distances in the UK are still measured in miles and, unlike mainland Europe, we remain wedded to mph on our speedometers.) When the train is approaching the park at a constant speed of 100 miles an hour you open the carriage window and throw the ball to your friend. From where you are standing on the moving train you are still throwing the ball at 2 miles an hour. Unfortunately, from where your friend is patiently waiting in the park, arms outstretched, this ball is now hurtling towards him at 102 miles an hour. Worse still, if the idiot attempts to catch it, he risks breaking a wrist and a couple of fingers.

There are several key points here. First, this is a really stupid thing to do. Second, and more importantly from a scientific viewpoint, the speed of the ball depends on the frame of reference and is relative to the location and motion of each person. It is travelling through the air at 2 miles an hour according to your position on the moving train and 102 miles an hour according to an observer on the ground – the friend with the throbbing hand. Third, we work out the speed of the ball from your, by now, ex-friend's point of view by adding the velocities (speed in a given direction) of the moving train and the thrown ball.

Let's repeat this experiment at night, this time replacing the ball with a torch. Don't worry, we're not going to throw the torch at your friend. We know that the speed of light is around 186,000 miles per second, or approximately 670 million miles per hour. When you are both standing in the park in the dark and you shine that torch, the light will travel from the torch to your friend at 670 million miles an hour, which, with both of you a short distance apart, appears virtually instantaneous. Providing no one has been arrested by now, if you then get on the train and shine the torch at your friend in the park as the train passes by, the inference is that something happens that is not supposed to happen. As the train whizzes past at 100 mph, the same velocity addition maths suggests that the light travels to your friend at 670,000,100 miles per hour – which means the speed of light just got faster.

If a voice in your head is saying 'Hello? I thought nothing could go faster than the speed of light,' congratulations. This is our problem. The speed of light is constant and cannot be exceeded. This means that the Newtonian method (after Sir Isaac Newton) of velocity addition for moving bodies must not apply for the speed of light, so there is a flaw in the theory. Enter Einstein.

Einstein made two important statements in his Special Theory of Relativity. First, the speed of light is constant and the same for all observers, no matter what their relative speeds. Second, the same laws of physics apply to each frame of reference (provided that the motion is uniform and there is no acceleration).

He also stipulated that there was no absolute frame of reference, so you cannot say that one frame of reference is the right one; and he assumed that the three dimensions of space and one of time are not separate but instead form a four-dimensional space-time continuum (an interpretation by one of Einstein's teachers, the German mathematician Hermann Minkowski). Incidentally, H. G. Wells also called time the fourth dimension in his 1895 novel *The Time Machine*, ten years before the formulation of Einstein's Special Theory of Relativity, thus setting a trend for many a science-fiction writer by predicting the future.

M&M – Michelson and Morley

The Polish-born American scientist Albert Michelson measured the speed of light in 1879 using an improved version of an experiment by J. B. L. Foucault (the French scientist who used the motion of a pendulum to prove that the Earth is turning). Like all good tricks, it was performed with mirrors.

Michelson placed two mirrors a long distance apart and rotated one of them so that when a beam of light was focused on the revolving mirror it would reflect to the other stationary mirror and back again. If you know the distance between the two mirrors, the speed at which the mirror rotates and the position of the reflected beam (its angle) when it arrives back at the mirror, you can work out the speed of light. Michelson got 186,355 miles per second. Amazingly close.

Eight years later, in 1887, Michelson collaborated with Edward Morley, an M&M partnership that produced unexpected results. The Michelson-Morley experiment was both a success and a failure. The aim was to detect the aether, which, at the time, scientists believed was some sort of transparent fluid that acted as a medium for light to travel through. To do this they decided to measure the speed of light relative to the aether, reasoning that, as the Earth was moving through this 'luminiferous aether' it would slow the light down slightly and this difference could be detected.

A beam of light was split in two, with each beam being sent via a separate route to a detector. Both paths were the same length. Michelson and Morley believed that if the Earth was moving through the aether, then the time taken to reach the detector by each light beam should differ. Sometimes the speed of light would be travelling through the aether in the same direction as the Earth orbiting around the Sun and so would be a bit faster; at other times it would be going in the opposite direction and so would be slightly slower. If this reasoning was correct, then the two light beams would produce black and white striped interference fringes of light.

It didn't quite work that way and so the experiment became proof that the aether did not exist. This conclusion was arrived at reluctantly because no one now knew what light was travelling relative to. The puzzle was answered by Einstein's Theory of Special Relativity, which declared that there was no rest frame of reference and light always travels at the same speed relative to the observer.

Space-time, being four-dimensional, needs four co-ordinates to describe it, in much the same way that you require four pieces of information to get to a party in a penthouse apartment (sounds more up-market than a tower block). To locate the street you effectively use two co-ordinates (latitude and longitude) to find it on a map. You then need a third co-ordinate in space upwards, corresponding to the floor number you have to take the lift to, and finally you need the fourth co-ordinate of time to make sure you arrive fashionably late.

Einstein then examined distance, time and the speed of light from two frames of reference: one on a moving train, the other on an embankment. He did this mathematically using all the above assumptions together with a set of equations known as the Lorentz transformation. His conclusion was startling: if the speed of light is to remain constant, then other things – such as time, distance and mass – must change. Time must vary depending on your frame of reference.

TIME TRAVEL FORWARDS

Einstein redefined time, saying that time is not absolute because it is not the same everywhere in the universe. Instead, both time and space are relative to where you are. From a potential time traveller's point of view, the most startling consequence of the accompanying mathematics is that if you approach the speed of light, time goes slow. This is also known as time dilation and produces the famous twin paradox.

Imagine twins, Sam and Samantha. One twin, Samantha, becomes an astronaut and heads off in a spaceship close to the speed of light (also referred to as a relativistic speed) to visit a planet in another solar system. Sam remains on Earth as he works for the council. When Samantha returns home, souvenir rock in hand, she will be in for a pleasant surprise. According to the Special Theory of Relativity, time has slowed down for Samantha from Sam's frame of reference and, depending on how many miles there are on the spaceship clock, she will be years younger than Sam on her return. Time is not the same from both frames of reference. There is no universal time.

Leaving on a jet plane

In 1971 scientists Joe Hafele and Richard Keating from the US Naval Observatory placed four atomic clocks on board a jet plane and flew them around the world twice. Once eastbound, once westbound. This is not so strange when you consider that Julian Lloyd Webber buys air tickets for his cello. Hafele and Keating were testing Einstein's prediction in his Special Theory of Relativity that time goes slow when approaching light speed.

There is one potential flaw here. Jets may be fast but they can't fly anywhere near the speed of light. Even so, the equations predicted that at jet speeds there would still be differences in time between a clock in motion and one on the ground. It was just that these time differences were so small they were only measurable in nanoseconds (billionths of a second), so to observe them required something a lot more sophisticated than a stopwatch. This is one of the reasons that Einstein's prediction took so long to verify. The only timepiece capable of measuring nanoseconds with the required accuracy for this experiment is an atomic clock, and in 1905 it hadn't been invented yet.

At the end of each trip, the times on the atomic clocks were compared with that of a reference atomic clock on the ground. Hey presto, time differences were found that matched those predicted by Einstein's equations. The Hafele-Keating experiment had worked. This is why.

Stand still and think about how fast you are moving. It's not a silly question. You may think you are stationary, but the Earth on which you are standing is rotating. At about 1,000 miles an hour if you're on the equator, slightly slower if you're elsewhere. See what we mean about frames of reference?

Now recall the throwing balls from a train analogy. It is the same principle for a jet plane. When a jet is travelling eastwards in the same direction as the Earth's rotation, its speed – relative to someone on board the jet – will be whatever the pilot says it is plus 1,000 miles an hour. If the jet travels the other way, westwards, opposite to the Earth's rotation, its speed will be less than 1,000 miles an hour. And, sure enough, the results showed that the atomic clock on the faster eastwards-bound jet slowed down in complete accordance with Einstein's predictions compared to the clock in the other jet and the one on the ground.

By slowing her ageing clock, Samantha has become a time traveller, because while visiting that new planet, she has also taken a trip into Earth's future. Unfortunately, this form of time travel into the future is limited, not least because it's a one-way ticket. It is also subject to interpretation. Samantha has not really travelled into her brother's future because Sam is simply 20 years older. There's also the small matter of building a spacecraft that can travel at near-light speeds. Yet, on paper, at least, travelling to the future (sort of) can now be done.

Proving the maths – that time goes slow when objects travel at high speeds – is another matter entirely. In 1905, when Einstein's ideas were published, the technology wasn't available to test his theory. Nearly three-quarters of a century later the world was ready and Einstein's time dilation equation was confirmed experimentally. Moving clocks go slow. Perhaps that's why the trains never run on time. Not in Japan, obviously.

Special relativity also predicts other changes in objects moving at close to light speeds. Not only does time go slow but mass increases and length contracts. So while Samantha gets younger during her space trip, from Sam's point of view on Earth, she also gets shorter if she's lying down in the direction of travel and thinner if she's standing up, and she weighs more. As far as Samantha is concerned, nothing has changed, and when she returns to Earth she will not look any thinner to us either, because she's no longer travelling at relativistic speeds. But as we're more interested in time travel, let's head off in another direction instead: general relativity.

GENERAL RELATIVITY

At first sight it does seem strange that Einstein's First Theory of Relativity is 'special' while his second – which took years longer to develop and is far more complex – is deemed plain old 'general'. Considering their impact, 'special' and 'extra special' might be more appropriate, but let's stick with convention and discuss the difference between Einstein's General Theory of Relativity compared to his earlier Special one.

The General Theory was published in 1916, although it was presented to scientists in its completed form in 1915, ten years after his Special Theory caused such a stir. The reason Special Relativity got its name was, remember, because it was for a special case: uniform motion only. General Relativity fills in the gaps. It applies to all those other cases where motion is not constant during acceleration or deceleration. Here's another difference. While Special Relativity considered space-time, General Relativity took those ideas and expanded them further to include the effect of gravity. This is why, among scientists, General Relativity is often referred to as Einstein's theory of gravity.

General Relativity says that gravity is not a force between two bodies, as Newton defined it, but results from warped, or curved, space-time. The best and most often quoted way to understand how this happens is to imagine a bowling ball on a rubber sheet. The bowling ball represents a large planet and the rubber sheet is a two-dimensional version of space. The heavy bowling ball naturally causes the rubber beneath it to sag. If you then add a tennis ball to that rubber sheet it will start to roll towards the bowling ball and will gradually pick up speed as it swirls nearer the ball because the mass of the bowling ball has warped, or curved, the space, or rubber sheet, around it. This gives the bowling ball an accompanying gravitational field, sometimes referred to as gravity well. Mass is therefore a source of gravity. The bigger the mass, the stronger the gravitational field.

Einstein also predicted that if there was a sudden change in space-time – say from an exploding star – gravitational waves would travel at the speed of light and transmit these space warps or distortions through the universe. No one has detected gravitational waves directly yet, but the search is ongoing. Perhaps his most surprising proposal, however, was that gravity slows time – an effect called gravitational time dilation. It took almost half a century for this to be shown to be true. Albert Einstein: he shoots, he scores.

It has been suggested that Einstein's genius was *how* he thought. He wasn't a truly great mathematician, by all accounts, yet he could see the big picture and envisage new possibilities for how things worked. It also helped that he had an active imagination. For Special Relativity he wondered how a beam of light would appear to moving observers from different frames of reference. For General Relativity he was considering how light would travel across a falling lift. As you do.

In 1907, in between the two relativity papers, Einstein made another important mental leap – one that effectively set up his Theory of General Relativity: he decided that the effect of gravity and acceleration would appear the same from the same reference point. This is called the Principle of Equivalence and Einstein described it as 'the happiest thought of my life'. Genius or not, he should have got out more.

GROUND FLOOR, PERFUMERY ...

Travelling inside a lift – not the falling one, we'll come to that in a moment – can help to explain what Einstein meant by his Principle of Equivalence. If you are inside a smoothly moving lift, with no windows to watch tell-tale signs of the world pass by, you've probably experienced that feeling where sometimes you can't tell if the lift is moving or not. It is only at the start of the journey when the lift accelerates and you feel a force (the sinking stomach sensation) or when the lift slows down (the rising stomach sensation) that you realise there is movement and you feel the effects of gravity. It either pulls you to the ground, making you feel heavier, or disappears when you slow down on the 500th floor, making you feel momentarily lighter.

In cheerful disaster movie mode, let's imagine a cable has snapped and that the lift is now hurtling to the bottom of the lift shaft. You, the sole passenger, are accelerating downwards at the same rate as the lift, floating around in free fall. Whereas at this stage most of us would be wondering when to start screaming or if weightlessness is really as much fun as it looks, Einstein imagined what would happen if a beam of light was shone across the lift during its descent.

If you could stop screaming for a moment and concentrate, you would see the light travelling in a straight line from one side of the lift to the other. On

The beam bender

Einstein proposed several ways to test his Theory of General Relativity. One of them was to measure the angle of starlight coming from a known star at night and then remeasure that angle during daylight, when the star's light has to travel past the Sun to reach the Earth.

According to Einstein, the Sun's large mass and its strong gravitational field would curve both space and time and bend the starlight, giving the star a different position in the sky relative to stars further away. His sums gave a specific deflection angle – twice that predicted by using Newton's theory of gravity as a force – so all that was needed was a suitable opportunity to put his conclusion to the test.

Luckily, the world didn't have to wait long. The only way to measure a star's position in daylight is during a total eclipse, when the moon completely covers the Sun and the ensuing darkness briefly allows stars to be seen during the day. In 1919, only a few years after the publication of Einstein's general paper, a total eclipse was predicted of around six minutes' duration, close to a cluster of stars in the constellation Taurus known as the Hyades.

British astronomer and physicist Sir Arthur Eddington organised two scientific expeditions to view this eclipse. One team went to Brazil, another – led by Eddington – to the Isle of Principe in west Africa. Each team was to measure the positions of the stars in the Hyades cluster close to the Sun during the eclipse. Astronomers already knew the position of these stars at night and so were waiting to see if these positions shifted during daylight and at what angle the Sun would deflect the starlight.

Einstein was right. The position of the stars had indeed shifted. According to his gravity theory (General Relativity), the Sun's giant mass and strong gravitational field warped both the space and time nearby and so the starlight travelled along this curved space-time on its way to Earth. This made headline news from the London *Times* to the *New York Times*, in which it was reported as 'Einstein's Theory Triumph'. Einstein's newfound celebrity prompted a number of articles and cartoons, as well as the following limerick by Arthur Buller – one that is well-known among scientists studying relativity:

There was a young lady named Bright

Who travelled much faster than light.

She started one day

In the relative way

And returned on the previous night.

At the turn of the twentieth century, Einstein's idea of gravity bending light would definitely have been called – if the phrase had been around at the time – thinking out of the box. Today's powerful telescopes allow astronomers to witness regularly the effects of gravity on light, as a gravitational field can also produce pretty patterns. Gravity can magnify or distort light into rings, arcs or multiple images – an effect called gravitational lensing. One particular case of gravitational lensing, in which a galaxy's gravitational field bends the light from an object behind it along four different paths into four separate images, is called Einstein's Cross in the great scientist's honour.

the ground it's a different matter. If a superhero with X-ray vision peered through that lift from a stationary position outside the building, the light would look as if it was doing something else to reach the opposite spot on the wall.

To our man in tights, the light would appear to be bending downwards. This makes sense when you consider that in the brief time the light takes to reach the opposite wall, the lift is also hurtling down the lift shaft and will have moved position slightly on its descent. To hit the exact opposite spot the light must bend to make its target. The Principle of Equivalence states that the effect of gravity and acceleration are the same, leading Einstein to reason that light must also be bent by a gravitational field.

Einstein predicted that light would bend across a gravitational field because this field was associated with curved space-time. But only something with a large mass and strong gravitational field – such as a star – would cause light to bend this way. For smaller mass bodies with weak gravitational fields Newton's theories still applied.

This theory of gravity produced a lot mind-bogglingly hard maths, known as coupled hyperbolic-elliptic non-linear partial differential equations. Fortunately, we don't have to deal with these in detail – that's for your degree course. All we need to know is that these formulae are often referred to collectively as Einstein's field equations and that the man himself admitted they were difficult. Mind you, Einstein didn't have to solve them – that would occupy other scientists for decades.

TIME TRAVEL BACKWARDS

We've seen how the twin paradox from Einstein's Special Theory of Relativity allows time travel into the future if someone travels close to the speed of light. As if that wasn't exciting enough, Einstein's General Theory of Relativity was found to open a door for time travel into the past. And if you want to turn back time and reverse all those mistakes you have made, this is the direction you will need to take.

In 1949 the mathematician Kurt Gödel found a new solution to Einstein's field equations involving a rotating universe, one that turns like a wheel round a central hub, and this also theoretically allowed time travel. The rotation didn't just apply to matter. It also applied to time and could lead to time loops. If you walked in the same direction as the rotation, then you would not only get back to where you started but you would also have travelled backwards in time. Einstein wasn't too pleased about this. From his point of view it exposed a potential weakness in his theory and it may have put a strain on his relationship with Gödel, because the Austrian logician was also his friend. No astronomer now believes that the universe is rotating – most think the universe is expanding – but Gödel showed the world that there were mathematical solutions that allowed time travel into the past.

Years later, in 1963, the New Zealander Roy Kerr put forward another solution for Einstein's equations: a rotating black hole. This, too, allowed time travel under certain circumstances. Since then there have been many other solutions allowing time travel using wormholes connecting two regions of space-time. The idea of time travel kept popping up and making a nuisance of itself or, depending on how you look at it, creating an exciting possibility.

In these solutions a journey along a timeline could take you backwards in time to arrive where you started. This is called a closed timelike curve – or CTC – a circular path through time where the start and finish points are the same. Wherever you are on the curved time loop, that position represents your past, present and future.

The 1993 film *Groundhog Day* is an entertaining way of seeing CTCs in action. Obnoxious television weatherman Phil Connors, played by Bill Murray, is in the town of Punxsutawney, Pennsylvania, for Groundhog Day. This is a rather quaint American tradition whereby the shadow of a groundhog determines when winter is about to end, but that's not important right now.

On the day of Connors's live report he wakes up at 6 am to the sounds of Sonny and Cher singing 'I've Got You, Babe' on a radio alarm clock. He showers, breakfasts, meets an old school chum and grumpily goes through the day, irritating his producer, cameraman and anyone else who meets him. That evening a snowstorm prevents everyone from leaving so he must, reluctantly, stay another night.

Next morning the radio alarm wakes him up at 6 am, again to the strains of 'I've Got You, Babe'. Believing that the radio station has mistakenly broadcast the previous day's taped programme, Connors heads for breakfast and discovers it's Groundhog Day. Again. Confused, he spends the rest of the day struggling to find out what has happened. A snowstorm prevents him from leaving. He spends another night in the hotel. At 6 am the alarm goes off and, yes, Sonny and Cher are singing 'I've Got You, Babe'.

Bill Murray's repeating sequence of events – in this case experiencing Groundhog Day every day – are those closed timelike curves, or CTCs, we talked about. Somehow, by 6 am each day, Murray has travelled back in time to where he started 24 hours earlier.

Groundhog Day is fun and doesn't pretend to be scientific – we never find out why a day repeats or why it stops – but there are some interesting points to be made from the film. When Connors goes back in time he does not meet another version of himself. Theoretically, he is 24 hours older on his first time loop so, logically, we might expect him to meet a version of himself that is 24 hours younger. He is also aware of what is happening and can change the immediate future, although only for 24 hours. If this had been a genuine CTC he should be either watching his past self do exactly what had happened the day before or be meeting multiple versions of himself. And, by the way, if Connors did meet another version of himself they wouldn't explode on shaking hands in a matter-antimatter annihilation kind of way, as often seen in sci-fi films. They could have a chat quite safely thank you very much.

A more scientifically accurate example of a CTC in the movies can be found in Terry Gilliam's film *12 Monkeys*. In it Bruce Willis plays James Cole, a criminal from the future who is promised freedom if he travels back in time to obtain a sample of a deadly virus that was released in the past, killing 5 billion people. The film flits back and forth at different points in history between 1918 and 2035, and we see a recurring dream of Cole's, in which a child watches a man being killed at an airport. The dream, we later realise, is an event from Cole's childhood. The man killed was Cole, back from the future and trying to secure the virus sample. As a child, Cole has therefore watched his own future death.

CAUSALITY

The key difference between these two films is causality. This is the relationship between cause and effect; what happens in the past affects the future. In *Groundhog Day* causality happens during the day. If the unpleasant weatherman stole money he got rich, but only for 24 hours. It is only in the final scenes, when he escapes the repeating closed time curve, that all he has achieved during his temporal imprisonment affects the outcome of his relationship with a television producer. By then he has learnt to play the piano and be kind to old ladies and is a thoroughly nice chap.

In *12 Monkeys* the implication at the end of the film couldn't be more different. Cole, as a child, has already witnessed his future, albeit unknowingly. He will be killed in an airport after travelling back in time on a mission. No matter what the time traveller tries to do while in the past, the conclusion is that these actions are already part of the past. In other words, the past cannot be changed. Cole will still die at the airport.

This is why most scientists believe that time machines can never be built and, even if they could, a time traveller would never be able to change the past: this would violate causality. Even if you went back in time and prevented your best friend from going to the beach on the day you knew he would die in a swimming accident, for example, your friend would die in a car crash instead. Despite altering the way in which your friend died, nothing ultimately would change because the death would still happen.

This train of thought is introduced into the otherwise extremely disappointing 2001 film remake of *The Time Machine* (directed, incidentally, by H. G. Wells's great grandson, Simon Wells). There are a number of inexcusable departures from the book – Morlocks who come out in daylight? But one addition, at least, adds a modern take on time travel into the past. After building his machine, the hero first goes back in time to try to prevent his fiancée's murder by a mugger in a park several years beforehand. This he happily manages to do, but while he is buying her some flowers she is killed in a road accident. He realises that while he can change the past by altering the way his fiancée dies, he cannot change the outcome: her death will always happen. To find out why, he must journey into the future.

There are other possibilities to explain causality of course. It could be argued that in *Groundhog Day* the weatherman entered a different parallel universe every 24 hours, as each day's outcome depended on which path he took.

GRANDFATHER PARADOX

The concept of parallel universes also provides some solutions to the grandfather paradox. The grandfather paradox is used to explain why time travel must be impossible. Imagine going back in time and somehow causing your grandfather to be killed before he meets your grandmother. This immediately creates a paradox. If your grandfather is dead, how could you have been born to go back in time in the first place?

This is where parallel universes can help. If a time traveller called Tim goes back in time and kills his grandfather, then there will be an alternative universe in which Tim never existed. There is also another reality in which the attempted murder never happened and both Tim and his grandfather are alive and well. This also ties in with the principle of self-consistency, which basically says that nature will not allow a paradox and that there will always be a self-consistent solution to a time travel problem.

In the film *Back to the Future* the teenage Marty McFly (played by Michael J. Fox) goes back in time and inadvertently upsets his own timeline by meeting a younger version of his mother in the 1950s. She starts to fall in love with him (there's a Greek play in there somewhere) and by introducing a new series of events into the past, Marty is in danger of preventing his future mother from meeting and falling in love with his father.

The grandfather paradox is addressed in the film in a simple visual way through a family photograph Marty has in his pocket. Whenever incidents threaten the future and increase the likelihood that his mother and father will never marry and have children, one of his siblings starts to fade and disappear from the photo. In this new future, Marty is in danger of never being born. Of course, Marty must have been born otherwise how could he have gone back to the past? But then that would have turned a full-length feature into a short.

WORMHOLES

A wormhole is another theoretical solution for Einstein's General Theory of Relativity. Needless to say, it means different things to a gardener than it does to a physicist. The wormholes scientists talk about have nothing to do with worms – unless those scientists happen to be gardening of course. Having said that, the physical description of a scientific wormhole is similar to that of the wormhole found in your average back garden. They are both cylindrical tunnels with an entrance and exit connecting one place to another. Where they differ is that one is real and found in soil; whereas the other does not officially exist and is a theoretical shortcut through space-time or, to excite the sci-fi fans, a portal to another world and potential time machine.

It all began when scientists were looking at black holes and discovered that, on paper, you could have a tunnel – or throat – connecting a black hole in two separate universes. This was termed an Einstein-Rosen bridge. Years later it was the explanation given for how, in the TV series *Sliders*, Quinn Mallory and his team could 'slide' from one parallel universe to another.

In the 1950s, physicist John Wheeler introduced the term 'wormhole' for any tunnel connecting two different regions of space-time. There are several ways to understand how a wormhole acts as a shortcut. If you've ever played the mystery board game Cluedo, you'll know that when trying to identify the murderer in the rooms of a stately home, a player can use a secret passageway to move from the conservatory to the ballroom. Instead of travelling square by square out of the conservatory, around the study and past the ballroom, a throw of the dice can achieve a much shorter direct route. A wormhole is a similar shortcut between two different places, but instead of a secret passageway beneath the rooms of an old house you travel through curved space-time.

A piece of paper serves as a good illustration of curving space to suit your interstellar travel needs. Grab a sheet of A4, place it on a flat surface and draw two small circles on it at opposite ends of the page. Cut the circles out and you have two holes representing planets Earth and Tharg. If you want to travel from Earth to Tharg you can get an idea of how long it will take you by looking at the distance between the two holes on the page. But if you now pick up that piece of paper and curve it, you create a shortcut whereby you can peak through the hole that is planet Earth and see the hole that is planet Tharg right next to it, a much shorter distance away. The connecting tunnel between these two holes in this form of curved space-time would be your wormhole.

No one has ever seen a naturally occurring wormhole, but John Wheeler believes that the space between atoms is filled with quantum foam containing microscopic bubbles and that these bubbles can produce wormholes. Unfortunately for potential time travellers, these wormholes would also be smaller than an atom. As if that wasn't disheartening enough, the quantum theory of gravity suggests that they would be also continually winking in and out of existence. Plus, wormholes are so unstable that they are liable to collapse before even a beam of light could pass through them.

CONTACT

Luckily there's nothing like a mathematical brick wall to get scientists thinking of ways to get round a problem. In the case of wormholes, inspiration came from an unusual source: science fiction.

Contact is a great book by astrophysicist Carl Sagan (published 1985) and an equally good film starring Jodie Foster as driven astronomer Eleanor Arroway, who searches for – and finds – evidence of extraterrestrial life. The book and film differ in a few respects but the basics remain the same. The story tells how scientists working for SETI (Search for Extra-terrestrial Intelligence) make first contact with an alien race.

Disturbingly, the communication is encoded within some black and white film footage of Adolf Hitler at the opening ceremony of the 1936 Olympic Games. This is because the Berlin Games were the first live television broadcast of a sports event in the world. As TV broadcast signals travel into space, they are eventually received by intelligent life across the universe. The rebroadcast was simply the aliens' RSVP.

The source of the signal is Vega – a star 26 light years away that would take hundreds of thousands of years to reach by conventional spacecraft. When the television signal is removed another message is discovered and, once decoded, is found to contain instructions for building a machine. In the film version, Dr

Arroway becomes the first human to test out the completed machine and discovers the equivalent of an underground tube network in space. It is, she speculates, some sort of wormhole system. It allows her to travel from Earth to Vega quickly and easily – somewhat like riding on a big dipper – with a few planetary tube stops along the way.

This form of interstellar travel was based on scientist Kip Thorne's re-examination of the maths for Einstein's Theory of General Relativity and his proposal of a stable wormhole as a feasible shortcut through space. Sagan, not being an expert on relativity or black holes, had asked Thorne, from California Institute of Technology (CalTech), if he would take a closer look at how his fictional characters travelled across the universe.

Sagan's original idea involved a black hole. However, as Thorne describes in his own book, *Black Holes and Time Warps*, he soon realised that Sagan's black hole would not work. Anything that entered it would be destroyed. It didn't take long for Thorne to realise what might work – a wormhole.

Helping out his friend Sagan set Thorne off in a new scientific direction, researching wormholes with his student Mike Morris. As a result of this research, published in a reputable science journal, came the suggestion that this same wormhole could also act as a time machine.

As we know, a microscopic wormhole would be of no use whatsoever to a five-foot-ten-inch would-be time traveller. So Thorne assumed that somehow an advanced civilisation could enlarge these microscopic wormholes for human use. But there was another major stumbling block. The wormhole had to be stable and, so far, all of the wormhole solutions for Einstein's field equations were unstable.

Thorne concluded that to keep the wormhole open there needed to be some sort of 'exotic' material to force the walls of the wormhole open. Exotic material is, as its name suggests, not your average kind of matter. It needs to have something called negative energy density (which basically means it weighs less than nothing) and negative gravity, so that it pushes outwards against the wormhole and stops you from being squashed as you travel through it.

In *Contact* it is proposed that an advanced civilisation constructed the wormhole transport system and so it is assumed that they are also able to use exotic material to keep it open. Sagan's main concern was for this to be theoretically possible, no matter how advanced and unobtainable the necessary technology might seem to us. A stabilised wormhole would allow safe travel across the vast distances of space in no time at all.

A couple of years after Sagan had originally approached Thorne regarding his interstellar transport system, Mike Morris was discussing wormhole travel with astrophysicist Tom Roman at a symposium in Chicago. Roman pointed out that the same wormhole could also be used to travel back in time. If a wormhole had an entrance and exit separated in time rather than space, you would get a time tunnel.

Roman's remarks together with Sagan's original request led Thorne, Morris and another (then) student, Ulvi Yurtsever, to produce their paper 'Wormholes, Time Machines, and the Weak Energy Condition' in the *Physical Review Letters* journal. This, together with the use of a wormhole in Sagan's *Contact*, led to the now popular use in science fiction of wormholes for time travel and fast interstellar travel across the universe or into a parallel universe. It also produced a sort of 'coming out' for several physicists, including Igor Novikov from the Nordic Institute for Theoretical Physics in Copenhagen. He was the man who introduced self-consistency, which basically says that the explanation that works will be the right one. It became acceptable (ish) to theorise about time travel.

Soon British physicist Stephen Hawking and other experts joined the debate. Hawking suggested that wormholes could be created but not for time travel because 'the laws of physics do not allow time machines'. To make his views clear, he proposed the chronology protection conjecture to make the world safe for historians. Besides, he famously argued, where are all the time tourists?

The response to this question is usually because a time machine hasn't been built yet. The logic behind this answer is straightforward. Until a time machine is built, no one can go back in time. This logic also assumes that you cannot travel back in time further than the date the time machine is built.

Both arguments appeared inviolate until, in March 2003, an American news story appeared on the web telling how a man had been arrested for insider trading. Andrew Carlssin reportedly turned $800 into $350 million in just two weeks with more than 100 high-risk trades. When Federal Investigators asked for the names of his accomplices, Carlssin claimed that there weren't any because he was a time traveller from 2256.

Our first time tourist appeared to have been caught, as there were no records of Carlssin to be found anywhere before December 2002. Although Carlssin offered information from the future such as the cure for AIDS and the whereabouts of Osama Bin Laden in exchange for his release, all those associated with the case reported being puzzled by his insistence that he was a time traveller. Needless to say, Hawking can relax; his chronology protection conjecture remains intact. The source of this wonderfully creative story was the *Weekly World News*, an 'entertainment' tabloid.

It's understandable that this story electronically circulated the world, because it buys into what most people immediately think of if asked what they would do if they could travel back in time. They tend to forget about righting wrongs, visiting loved ones and retaking exams; instead they are preoccupied with getting rich. Imagine travelling back to a time when a house cost half a crown and buying property with a few old bank notes. Or playing the stock market, as the imaginary Carlssin did, but a little more discreetly. Perhaps a series of cash bets on those 300:1 shots at the races is a far better way to make instant and untraceable millions. Or travelling back in time to get on the TV programme *Who Wants to Be a Millionaire?* knowing all the answers. You'd just have to make sure you didn't arouse suspicions by not reasoning your answers through on air – or by taking along an accomplice along with a bad cough. We all know where that sort of behaviour can lead to. In Major Charles Ingram's case, a suspended jail sentence and no money. At least not from the programme.

Unfortunately, no one at Ingram's trial thought to ask if he had built a time machine and travelled into the past. If they had, perhaps more pressure would have been put on Hawking's chronology protection conjecture. At the moment, though, it stands. Nevertheless, Hawking has changed his mind about time travel and now believes that a wormhole time machine might be possible if there are parallel universes. But if we are going to travel through any wormhole we are definitely going to need some of that exotic material.

ANTIGRAVITY AND NEGATIVE ENERGY

If our best bet for time travel involves creating a wormhole and stuffing it full of Thorne's exotic material to keep it open, then we are going to have to go shopping. Exotic material, as its name suggests, is not your ordinary material, and we need some that has negative gravity, or antigravity. The very word 'antigravity' can produce apoplexy in some physicists (and do read Nick Cook's book *The Hunt for Zero Point* if you want an entertaining account of the conspiracy theories and history surrounding the search for antigravity), but it has been produced in a laboratory.

To produce negative gravity we need negative energy, which means the energy level will be less than zero too. Although negative energy will sound familiar to anyone who has struggled to get out of bed in the morning, on a scientific level it is a much rarer beast. It was once considered impossible to achieve by many scientists, until Dutch physicist Hendrik Casimir proved that it existed.

Casimir and D. Polder predicted in 1948 that if you placed two metal plates face to face in a vacuum (a box of empty space) they would be attracted towards each other. It was a bizarre suggestion and yet it is indeed what happens at extremely small distances. If the plates are billionths of a metre apart they are seen to move together. The force between the metal plates is called the Casimir force, and the closer the plates, the larger the force.

It happens because, even with the air sucked out, the empty space in a vacuum is not empty space at all. It contains 'virtual' particles. These result from Heisenberg's Uncertainty Principle, which states that you can never completely determine a particle's position and its momentum at the same time. On a quantum level, even in empty space, there will therefore be loads of particles blinking in and out of existence. Because they pop in and out of reality, these are called virtual particles.

As particles can also be described as waves, different particles have different wavelengths, so not all of these virtual particles will be able to squeeze through the gap between the two plates. In much the same way that a medium-sized body can't fit inside a small-sized pair of trousers (no matter how hard you try), particles whose wavelengths are longer than the distance between the plates can't fit between them. This leads to an inequality between the number of particles outside the plates and those inside. Therefore if there is zero energy inside the vacuum then the energy between the plates must be lower than that surrounding it and so will be less than zero, or negative. This negative energy will then cause the plates to move together, which is exactly what happened in Casimir and Polder's experiment.

This negative energy is, as far as we are concerned, exotic enough to qualify as the sort of stuff we need to produce negative gravity and keep our wormhole open. So far only small amounts of negative energy have been produced in a laboratory, and scientists have speculated about how to make it on a much larger scale, not least because calculations show that about a planet's worth of negative energy would be required to build a wormhole only 1 metre across. Not just any old planet either. The one they have in mind is Jupiter – the largest planet in our solar system.

THESE ARE A FEW OF MY FAVOURITE STRINGS

String theory is unusual enough to be called exotic but it has nothing to do with exotic matter. It is based on imagining particles not as single points but as tiny closed loops of string. Some physicists believe life can be explained by a string version of a spaghetti hoop.

There's more. The loops that make up the particles are vibrating, in the same way as a piano or violin string, and each is capable of different modes of

Superstrings

Superstring theory attempts to explain the fundamental processes in nature. It is one theory, as opposed to several, to account for why everything behaves the way it does.

String theorists believe that superstrings unify or plug the hole between relativity and quantum theory. The hole is there because of a problem with size. Quantum theory works on an extremely small scale and describes why atomic particles behave the way they do. Einstein's Theory of Gravity applies to large objects such as planets and stars. Everything works out fine when the two theories are used separately, but there are certain events in cosmology that need to be examined by both theories at the same time.

Take a cosmic black hole, for instance, with a mass at least tens of times greater than that of our own Sun. As it has a large mass, scientists need to apply the Theory of Gravity. Yet they also need quantum theory to explain what is happening at the central point. It is when both theories are combined that, mathematically, bad things happen. They are incompatible.

String theory resolves this incompatibility. When a particle is considered as a sub-atomic spaghetti hoop, string theory agrees with Einstein's Theory of General Relativity at large distances and can replace it at extremely small distances (where normally quantum theory would take over).

String theories have been around in some form or other since the 1960s and a number of scientists helped formulate them. Superstring theory arrived in 1981, with John Schwarz from the California Institute of Technology and Cambridge University's Mike Green being the prime movers. Schwartz and Green discovered that superstring theory required one rather special added extra. For all the sums to work, we need ten space-time dimensions rather than four – one of time and nine dimensions of space. If you are wondering why we've never seen these extra dimensions, the theory also predicts that after the Big Bang only four dimensions expanded and the other six curled up into a microscopic ball. Life makes you do that sometimes.

Edward Witten is one of string theory's pioneers. Based at Princeton University's Institute for Advanced Study (located on Einstein Drive, trivia fans), Witten showed, in the mid-1990s, how the different string theories were all part of the same grander scheme. He called his discovery M theory. It is an 11-dimensional theory in which, as well as strings, there are also membranes with up to ten dimensions in space (11 in total if you add time). These different membranes are called p-branes, where p is a number from 0–9. A point is a p-brane of zero. A string is a p-brane of 1. A bubble is a p-brane of 2. M theory is also referred to as 'the theory formerly known as string', and debate surrounds what M stands for. Some say magic, others 'mother of all theories'. We prefer mind-blowing.

vibration or harmonics. This has led to an endearingly poetic description of the universe as a symphony of vibrating strings. It's unlikely you could hear this symphony, however, as the average length of each string is a millionth of a billionth of a billionth of a billionth of a centimetre. The vibration of the string also accounts for a particle's charge and mass and for the four fundamental forces of nature – the strong nuclear force (which holds everything in an atomic nucleus together), the weak nuclear force, electromagnetism and gravity.

There are basically five different string theories. One favours open strings (spaghetti – no hoops) while others favour closed strings (spaghetti hoops) that spin in a particular direction. This is a hot area of theoretical research but there is no proof for any of these theories. If strings do exist, the reason proposed for why we've never seen them is that we haven't got the equipment to give a close enough view of particles yet.

Some scientists are working on superstring theory (which kind of combines all the others) as the 'theory of everything' – the theory that will unify all existing theories and explain what is happening at all scales, incorporating both general relativity and quantum theory. If it works, this will help scientists stabilise our theoretical wormhole time machine. But it's no easy task. Einstein spent the last years of his life obsessively trying, and failing, to come up with a theory to unite gravity and electromagnetism. No wonder the 'theory of everything' has now reached the status of a scientific Holy Grail.

HOW TO BUILD A TIME MACHINE

In fiction it is hard to beat the Victorian contraption built by the time traveller in H. G. Wells's 1895 novel *The Time Machine*. Although not much detail is given about its construction in the book, the description of materials – brass, ivory and quartz – suggest a machine of engineered beauty and elegance reminiscent of a grandfather clock or an early automobile.

The 1960s television series *The Time Tunnel* had a time machine straight from a science-fiction fan's imagination. It was a tunnel of concentric black and white circles appearing to extend to infinity, with some atmospheric blue smoke in the distance. Groovy. Built as part of the US government's Project Tic Toc, the Time Tunnel has trapped two scientists, both of whom have got into historical scrapes, watched by their helpless colleagues back home in the time tunnel lab. No attention whatsoever is paid to any time travel paradoxes, but who cares? It's great fun. This monochromatic sixties-style Time Tunnel makes a memorable re-versioned reappearance in the Austin Powers movies when actor Mike Myers's international man of mystery goes back in time to save the world in *The Spy Who Shagged Me* and *Goldmember*.

Dr Who's time machine, the TARDIS, was, famously, an old blue police telephone box that could travel through time and space. 'TARDIS' stands for Time and Relative Dimensions in Space, and Dr Who was a time lord (as if you didn't know). His TARDIS, like those belonging to other time lords, was supposed to change its appearance on materialisation so that it blended into the surroundings. Conveniently for the limited BBC budget, Dr Who's TARDIS was broken and always reappeared as a blue phone box.

A telephone box time machine pops up again in the film *Bill and Ted's Excellent Adventure* (1989). A young Keanu Reeves is one of two student dudes who use a modern phone box to travel through time collecting famous historical figures to help them with their homework. Don't ask why, it's way too silly. Their phone box didn't have the same internal dimensional capacities at the TARDIS, so it was quite a squeeze bringing back the likes of Napoleon and Genghis Khan.

For sheer fun, it's hard to beat the inspired time-travelling device used in the *Back to the Future* movies: a souped up DeLorean car with that all-important flux capacitor. All you need is a mad scientist (naturally) and enough juice to hit 88 mph.

More recently, several physicists have come up with blueprints for building a time machine. Physicist Jim al-Khalili covers this in his book *Black Holes, Wormholes and Time Machines*, while Paul Davies has written an entire (and extremely entertaining) book on the subject called *How to Build a Time Machine*. Unfortunately, unlike H. G. Wells's time machine, this is definitely not a home do-it-yourself project, not least because it effectively involves as a starting point, take one wormhole ...

Davies explains an ingenious method that involves stabilising a wormhole with negative energy so that it doesn't become a black hole, using an advanced technology spacecraft to separate the entrance and exit and then towing one end of this wormhole with an electric field to a neutron star. The strong gravity of the neutron star would cause time to slow down at one end of the wormhole. Time at the entrance and exit of the wormhole is now happening at a different rate. If you went in one end you would emerge in the future; if you went in the other end, the one where time has slowed down, you would emerge in the past. The wormhole is now a time machine. No one said this was going to be easy.

Despite all the fun scientists have thinking up ways to make a time machine from a wormhole, there's one crucial problem: no one has ever seen a wormhole and there is no evidence that they exist. The good news is that no one has been able to prove that they do not exist either. Given that the history of science is filled with theoretical predictions that take years – sometimes decades – to verify experimentally, nothing can be ruled out for potential time travellers.

COSMIC STRING

OK, you can handle the concept of using brass, ivory and wormholes to build a time machine, so how about some string? Not just any old string. Cosmic string, to be precise, and it is not the same type of string we discussed earlier. Cosmic string is astronomically long and is thought to be strands of high-density matter left over from around the time of the Big Bang. Some people describe it as creases in our universe, although these creases contain so much mass that a kilometre of cosmic string would weigh as much as the Earth.

Cosmic string was first proposed in 1985 by J. Richard Gott and William Hiscock and – like wormholes – are a theoretical solution to Einstein's General Theory of Relativity. Considered relics from the birth of the universe, cosmic strings are thought to contain exotic matter.

In 1991, Gott proposed a time machine made from cosmic string in the journal *Physical Review Letters*. Needless to say, it's not easy. It means finding and twanging a large loop of cosmic string so that it contracts. Due to the string's massive energy density, the contractions will warp space-time. If two sides of the cosmic string loop are then travelling past each other in parallel and close to the speed of light, then if you fly a spacecraft around the loop at the same time, you will travel back in time. If you haven't been torn apart by the curvature of space of course.

TURNING BACK TIME

There are enormous hurdles to overcome before we are able to build a time machine and change the course of our present life. Cosmic strings and wormholes have to be proven to exist for a start. Even if you are only interested in travelling to the future, there's the technological advance needed to send spacecraft at speeds close to the speed of light. It is also, remember, a one-way trip.

If the theoretically predicted wormholes are ever found, there is then the matter of expanding these microscopic tunnels sufficiently large enough for them to accommodate a human being or a spacecraft and stabilising them so that they don't suddenly shut and cut off your journey – and everything else – in its prime. Exotic matter will be needed in quantities far greater than is currently achievable in a laboratory. Also, for the type of wormhole time machine suggested by Paul Davies, the running costs are literally astronomical, not least because it requires a neutron star.

Some scientists hope that, because wormholes are predicted mathematically but seem so enormously difficult to build, nature will do it for us. Wormholes, fingers crossed, will occur in space naturally. It's just up to us to discover them somewhere in the universe, already up and running and ready for volunteer time travellers.

Let's recap on what we have here. Travelling back in time requires all or most of the following:

- ❀ A wormhole (which has yet to be proved to exist)
- ❀ Cosmic string (ditto)
- ❀ Exotic matter (currently unavailable in the large quantities required)
- ❀ A neutron star (currently unavailable in any nearby stores)
- ❀ A spaceship that can travel close to the speed of light (only available in science fiction)

At this stage it is tempting to think that maybe Jack Finney got it right in his book *Time and Again*. Finney's hero travels back in time through self-hypnosis, which is so much simpler. Or perhaps we should dismiss the idea of time machines as something that will never happen. Except history is filled with predictions of impossibility that are, in hindsight, so off the mark as to make us feel unbelievably smug and technologically superior. There are hundreds of quotes to choose from on this theme, but let's single out the English scientist William Thompson who, in 1895, was president of the Royal Society, Britain's most prestigious scientific organisation. 'Radio has no future,' he said. 'Heavier-than-air flying machines are impossible. X-rays will prove to be a hoax.'

Exactly. So remember this: despite the seemingly insurmountable physical difficulties in building a time machine, there remains nothing in the laws of physics to prevent time travel. If the law of causality is not to be violated, however, there will be nothing you can do to change any event in this world. The most promising option, in our opinion, is to travel back in time into a theoretical parallel universe and live out the life you've always wanted. Unless you are one of the lucky ones, of course, who is already living the life you've always wanted in this world.

CHAPTER SIX

How to Upgrade Your Body

Get tired just carrying the shopping? Out of breath when running for the bus? Can't remember phone numbers, let alone the date of your mother's birthday? Can a nine-year-old defeat you at arm wrestling? If you answered yes to any of the above, you need some silicon enhancement. A body upgrade. After all, what's the point of losing those love handles if other parts of the body let you down.

When it comes to the human body, technology can do many things so much better. It has the potential to give you strength, knowledge and stamina. It can improve your vision, hearing or smell. We're not suggesting giving up your body altogether but merging it with some technological attributes. Becoming a cyborg – part human, part machine – holds limitless possibilities. Better, stronger, faster, as it says on the packet. And the good news is you can probably do it for a lot less than 6 million dollars.

Science-fiction writers are extremely fond of cyborgs, from Daleks to Darth Vader, bionic men to Borgs. We're not suggesting you'd want to spend the rest of your life shuffling around in a box with a sink plunger, but cyborgs certainly hold no end of exciting possibilities. Not only could the Six Million Dollar Man lift cars, see vast distances and leap buildings in a single stride (or was that Superman?) he could run in slow motion too. How cool is that?

OK, Daleks, Darth Vader and the Borg are also some of sci-fi's most enduringly (and endearingly) evil characters but it doesn't have to be that way.

Utilising cyborg technology doesn't necessarily mean you renounce humanity and become bent on world domination. Although it would probably help.

So how can it work for you? Imagine for a moment that you are somehow wired up to a computer, just an ordinary PC, and that that computer is connected to the internet. You now have access to an incredible amount of information – from the telephone number of your dentist to facts like the average distance of the Earth from the moon (384,500 kilometres in case you were wondering). You know the spelling of every word, can check the latest news or sports results, watch movies or listen to music. All in your head. You are hard-wired to the world. Then imagine every other person has got that connection as well. Now you can communicate with anyone, anywhere, whenever you want, holding conversations in your mind with people thousands of miles away.

What about your body? You'll probably be the first to admit you've been letting yourself go recently. Rather than spend a fortune joining a gym that you'll lose interest in after a couple of weeks, why not pay for some serious surgery instead? Then you can replace those worn-out knees with new ones, those muscles with motors. You could even replace your whole body with a new mechanical one.

This is not as far-fetched as it sounds, because it you want to become a cyborg, much of the technology already exists. Cyborgs are already walking among us. Anyone with a pacemaker or artificial hip incorporates technology directly into their body. One of the most sophisticated cybernetic devices currently on the market is the cochlea implant to help you hear. Yes, your granny is a cyborg.

LENDING AN EAR

The size of a frozen pea, the cochlea is part of the inner ear. It converts sound waves entering our ears into nerve impulses for our brain to interpret. The cochlea is the human equivalent of a microphone, transforming sound into electricity. For people with normal hearing, soundwaves travel along the ear canal and cause the eardrum to vibrate. The three smallest bones in the body, the malleus, incus and stapes (which are also known as the hammer, anvil and stirrup because that's what they look like), then amplify the sound. The resulting vibrations pass through the oval window (never the square or round window, we're afraid) to the inner ear.

The cochlea itself looks a bit like the shell of a snail (in fact *kochlias* is Greek for 'snail'). It's divided up into spiral cavities filled with fluid and separated by bony walls lined with tiny hairs. Once the sound wave arrives at the inner ear, it becomes a pressure wave in the fluid. Different parts of the cochlea respond to different frequencies of sound. So only certain auditory hairs receive each particular 'note'. Within the hair cells a chemical reaction takes place generating an electrical current. This is transmitted to neighbouring nerve cells, where it can be fed into the brain.

The whole mechanism is incredibly fragile, particularly the hair cells, and not surprisingly it's prone to failure. So in the 1960s a team of Australian scientists, led by Professor Graeme Clark, at the University of Melbourne, developed the cochlea implant. It replaces not just the cochlea but almost the entire ear with electronics.

Anyone with an electronic ear 'hears' through a tiny microphone that converts sound waves into electrical waves. This is connected to a speech processor, which generates corresponding electronic signals (a series of pulses). These are transmitted to electrodes in the cochlea, which stimulate the nerve cells.

Since the first artificial cochlea was implanted in 1978, the electronic technology has got smaller and more sensitive. Tens of thousands of people have now had the operation, many of whom had lost their hearing altogether. At the moment an implant is still not as good as a fully functioning human ear, but reaching that goal is probably only a matter of time. Rather than stimulating nerve cells in the cochlea, a future implant might attach directly to the brain.

The key to the success of the cochlea implant is the actual interface between the human and the machine, and that's true of any cyborg technology. The cochlea implant is not directly wired into the nervous system, but is able to stimulate nerves in the right way to produce something the brain can deal with. Likewise, a Dalek has got to have some sort of direct or indirect connection to its sink plunger. A Dalek might well have the most sophisticated sink plunger in the galaxy and a mind of pure logical evil, but if it can't combine the two together, then the plumbing business goes down the drain.

Some writers, philosophers and scientists believe we will never be properly integrated with machines until we undertake some serious surgery. However, there are others who argue that we have already been assimilated.

MAKE MY DAY, CYBER PUNK

We interact with technology every moment of the day – from the alarm we grope for first thing in the morning to the computers we scream at when they decide our work is no longer necessary to the sum of human greatness. It's these interactions with technology, the argument goes, that make us cyborgs. The most widely quoted exponent of this belief is a Californian academic and feminist writer, Donna Haraway, author of *A Manifesto for Cyborgs*. She has been a major influence on cyber punk writers and has surely got a point. What difference does it make whether we are connected directly to the computer or whether we interact with it through our hands or eyes?

As technology improves, so does the human interface with it – user-friendliness is inherent in the cyborg dream. Computers are a good example of how that human–machine relationship has changed. Despite our daily frustrations with products from the Bill Gates empire, the PCs we use today are, believe it or not, designed with us, the user, in mind.

To type a sentence on a page, for instance, you do not have to understand how your computer works or how it is translating your key presses into words on the screen. You just have to know how to switch the thing on and remember where you saved the document you want to work on. We take a lot of this for granted, but even 20 years ago things were very different. The earliest computers didn't have displays or keyboards. The data was fed in on punch-cards and spewed out on ticker tape. Computers were designed to be used by the sorts of people who knew how they worked, such as scientists and engineers. They were machines created to solve complex equations or plot missile trajectories. It was only when 'normal' people started using computers that the concept of user-friendliness was contemplated.

One of the largest computer systems ever built was the SAGE (standing for Semi-automatic Ground Environment) network for the United States military. An early warning system, it was designed in the 1950s to co-ordinate radar stations across the North American continent. If an enemy plane or missile were spotted, the network was designed to trigger the launch of interceptor aircraft. The original goal was to make the whole thing automatic. Spot a plane, shoot it down. Perhaps understandably, this was deemed a little rash. Instead the system relied on human operators to make the key decisions. These were not highly trained computer geeks, but soldiers.

The consoles they used were some of the earliest to incorporate user-friendly elements. The operator's interaction with the computer was through simple controls and light pens (electronic pens that they could point at the screen). The display was a visual representation of the area they were responsible for, not a load of computer code. As it turns out, historians have concluded that the SAGE system was a complete waste of time and money, being largely a public relations exercise to convince the American public that they were safe from commie bombs. But the vast expenditure the US military threw at the project had the unexpected side-effect of helping to break down the barriers between humans and computers.

An American scientist, Douglas Engelbart, came up with the idea of using a mouse to help us navigate around a computer system, manipulate files and text, and run programs. Instead of lots of words, the main interface of a modern PC now uses icons as representations of particular things we might want the machine to do, which makes it all easier to understand and interact with. We are using the computer as an extension of our body.

IS THAT A PDA IN YOUR POCKET OR ARE YOU JUST PLEASED TO SEE ME?

The Personal Digital Assistant (Palm or pocket PC, among others) has given that human–computer interface portability. We can check appointments, find addresses and look up phone numbers. Connected to a mobile phone, we can access the web and a wealth of information and entertainment. To do this we have to use our eyes and our fingers, but does that make us any less a cyborg?

Rather than sticking the computer in our pocket, how about incorporating it into our clothing? A wearable computer could include all the advantages of a pocket PC but without the pocket. One of the pioneers of what are called 'smart clothes' is Professor Steve Mann, who started attaching computers to himself at the Massachusetts Institute of Technology (MIT) Media Lab. By incorporating the visual display in a pair of glasses, he can go about his normal life while referring to his computer. The glasses contain a camera, and the computer a modem connecting him to the worldwide web. The design has given him an enhanced view of the world and gives the rest of the world a view of what he is seeing.

Smart glasses are a variation on the technology used by the military, only don't involve killing people. 'Heads-up' displays project information for fighter pilots right in front of their eyes. They enhance the pilot's view of the world with information about terrain, targets, fuel, etc. so that he or she doesn't have to look down at the controls and can concentrate on flying the plane.

The glasses worn by Steve Mann and the helmets worn by pilots will not win any awards in the fashion stakes (unless you're into that sort of thing) but with the increasing miniaturisation of technology, an ordinary pair of glasses could incorporate many of the same features. While laser targeting may not be your first priority, it might be useful to have a map superimposed in front of your eyes as you navigate an unfamiliar city. By connecting your glasses to the web you could not only access a world of information but also receive emails or send images of what you're seeing to friends. Of course, you need to remember to switch it off when you go to the toilet.

One of the latest projects at the MIT Media Lab aims to give these devices memories. By incorporating the advantages of a diary, or PDA, they are designing 'memory glasses' that will remind you of meetings, directions or anniversaries. The key to their development is making the glasses 'context aware' so they only deliver information at appropriate times, otherwise they might deliver football scores while you're crossing a busy road or in the middle of a job interview. A similar concept was employed by the late Douglas Adams in *The Hitchhiker's Guide to the Galaxy*, in which one of the heroes sports 'peril sensitive' sunglasses that turn completely black at signs of danger – thus making the wearer blissfully unaware of the thing they should be frightened of.

By incorporating artificial intelligence into the glasses, other researchers are attempting to build in face recognition software. Every time you look at someone their identity will flash up in front of your eyes. A real boon if you have difficulty remembering names and an essential piece of kit for the robo-cops of tomorrow. By giving the glasses some 'intelligence' scientists are, in effect, enhancing the user's own.

SMARTY PANTS

Another idea from the MIT labs is smart underwear – and not just for those who keep their brains down there either. Knowledgeable knickers have practical applications that could make life more comfortable. When wired up to sensors and a transmitter they can be used to control the temperature in a room, depending on a 'sweatiness index'. The more sweat that is generated, the more the room temperature is turned down. Unlike a conventional thermostat, smart knickers adjust the room temperature to the individual. This isn't as barmy as it sounds. Perhaps you have just taken some exercise and are too hot; the pants will sense this and turn down the heating. Of course, it's fine if you live on your own, but if you share a house and both wear a sensor, deciding whose pants control the temperature could be a new source of domestic disharmony. The idea also brings a new meaning to who wears the trousers in a relationship.

Although most of these ideas are at the prototype stage, some are under serious consideration by major electronics manufacturers. The big drawback at the moment is that they don't look so good. They are all electric, so they need batteries; they are wireless, so they need transmitters. If you want your underpants to look sexy as well as smart, then you've probably got a long time to wait.

Although wearable computers undoubtedly hold possibilities, they do not quite make you bionic. Memory glasses still rely on eyes to see; you can't *think* information onto the screen. Take your smart pants off and you are no longer superhuman.

CAPTAIN CYBORG

It's a crisp winter's day in 2002. At a secret location somewhere in Britain, a university professor lies on an operating table, his arm outstretched. A microscope is positioned over his wrist, the image displayed in colour on two large monitors slung from the ceiling. A team of surgeons, nurses and technicians, all dressed in light blue gowns, busy themselves around him, talking in hushed, calm voices. Kevin Warwick is having a chip connected directly to his nervous system to make himself a cyborg.

Warwick isn't the shy and retiring sort. It's no coincidence that this surgery is closely linked to the publication of his latest book. Neither is he without his critics. There is something of a small industry of spoof websites dedicated to the exploits of 'Captain Cyborg'. But say what you like about him, Kevin Warwick is nothing if not brave.

We were the only two journalists allowed in the operating theatre for the somewhat gory two-hour procedure. The surgery involved attaching a silicon chip around a quarter the size of a penny coin onto the median nerve of Warwick's wrist. When the skin was cut open, the nerve itself was clearly visible as a thin white tube just above the bone. The median nerve controls most of the movement and feeling in the hand, so any mistakes and Warwick could have lost its use altogether. On one side the chip was covered in an array of tiny electrodes which were stamped into place on the nerve using what amounted to a surgical stapler. Wires from the silicon implant were passed under the skin to emerge just below the elbow. When the wound was sown up all that was visible were some wires protruding from the skin attached to a circuit board. And a lot of exciting multicoloured bruising.

There was a serious intent behind the operation – and, indeed, had it been merely a publicity stunt, no responsible surgeon would have performed it. If the procedure proved successful it might help people with disabilities regain movement in damaged limbs or help amputees better control prosthetic arms or legs.

A few days after the operation, Professor Warwick was attached to a robotic metallic hand. He found he could position the robot by moving his own wrist. The chip was clearly working. The electrodes imbedded in his median nerve were picking up the electrical impulses coursing through his nervous system. These pulses were clear enough to make sense to a computer and be converted into the appropriate commands to operate a machine. He tried the same thing across the internet, controlling the robot in Reading, England, over a phone line from New York.

Although the experiment is relatively simple, the concept is impressive. Warwick's nervous system was controlling a machine thousands of miles away. Cyborgian technology could allow us to control devices on the other side of the world through thought alone. We could turn on the dishwasher, the heating or the lights at home. Think how much easier it would be to set the video. No longer flummoxed by the controls, we could just *think* it into recording *Buffy*.

There was another clever thing about Kevin Warwick's implant: it worked both ways. The chip could stimulate the nerve, making the professor's fingers twitch. However, it was not particularly precise. The median nerve consists of thousands of nerve fibres. Punching electrodes into them stimulates the whole bundle, not individual fibres, making delicate control largely impossible – nevertheless, that it worked at all is encouraging. Imagine what could be achieved with a direct connection to the brain. That's exactly what other scientists are attempting to do.

THE SIX-MILLION-DOLLAR EEL

The sea lamprey isn't one of nature's most attractive creatures. At approximately 40 centimetres long, the eel-like animal is one of the most primitive fishes on the planet. Around since before the dinosaurs, sea lampreys attach their jaws onto other fish and suck on their blood and bodily fluids. Which is nice. They are a relatively simple animal and their nervous system has been extensively studied. This makes them an ideal candidate for cyborg technology. In May 2000, scientists at Northwestern University Medical School in Chicago revealed that they had connected the brain of a sea lamprey to a small robot. The fish could now control the machine.

To achieve their eelbot the researchers, led by Professor Sandro Mussa-Ivaldi, had first to remove the sea lamprey's brain stem and some of its connecting spinal cord. They found they could keep the whole lot alive by preserving it in cold salt solution, which mimicked the seawater the fish normally lived in. They connected electrodes to the nerve cells that normally process information from the sea lamprey's sensory organs, and other electrodes were attached to nerve cells that control movement.

Normally, when a sea lamprey swims through water it uses sensory cells to help with orientation. Without them it wouldn't know whether it was swimming upside down. In the eelbot, these were replaced by light sensors on the robot. Signals from these electronic sensors were processed and sent to the disembodied brain stem, which responded with nerve impulses to control movement. These were transmitted back to the robot and converted into appropriate controls for motors driving the robot's wheels. Depending on where the researchers placed the electrodes, signals from the fish brain sent the robot either towards or away from the light. If we assume that the sea lamprey is not truly conscious, as far as it was concerned it was still swimming in the sea as the robot spun around on the lab bench.

Obviously, a bionic eel isn't much use to anyone but it certainly demonstrates the possibilities. The nerve cells of a fish are fundamentally the same as those of higher animals, so if it works for them, there's no reason why it shouldn't work for us. But probably best to try it on something else first.

CYBER MONKEY

In the same year that the lamprey got wheels, research was published in the respected science journal *Nature* demonstrating that a monkey could control a

The real man with two brains

If you have been feeling a little uncomfortable with the ethics of some of the experiments we've been talking about, look away now. Because it gets far worse.

In what is possibly the ultimate attempt to enhance the body, an American researcher transplanted the whole head of a monkey onto the body of another. The procedure was carried out by Professor Robert J. White, from Cleveland, Ohio – and described by British biologist Professor Stephen Rose as 'wholly unethical and inappropriate for any reason'.

If the idea sounds familiar, it's employed to brilliant effect in the Steve Martin film, *The Man with Two Brains*, in which Martin plays the inventor of screw-top, zip-lock brain surgery. The film is memorably funny less for the brain transplant antics (which involve mad scientists, an over-sexed surgeon and Kathleen Turner) and more for the constantly mispronounced hero's name: Dr Michael Hfuhruhurr.

Back in the equally bizarre real world, White explained how he had been able to transplant his monkey brain into an 'intact' animal and keep it alive for several days. What he had been unable to do was to connect up the two nervous systems. The body of one monkey was keeping the head of the other alive.

Impressive? Maybe. Grotesque? Certainly. Useful? It depends. White argues that experiments like this will make the next step in human organ transplants possible. For example, one person might be brain-dead but have a perfectly serviceable body. In normal circumstances the corpse might well be used for organ transplants anyway (the heart, kidneys, liver, etc. could all be removed). In White's scenario, rather than just taking the parts, the whole body could be utilised instead. Potential recipients would be those with serious spinal injuries who have lost the use of and feeling in every part of their body below the neck.

Walking around in someone else's body raises so many ethical dilemmas that it's a book in itself. How are the friends of the person whose body it was going to feel? What about their wife or partner? Could people be killed for their bodies? Critics also argue that it would be far better to invest research efforts in studies on the regrowth of nerves to restore movement and feeling after serious injury.

You could, of course, go down the route of Zaphod Beeblebrox in *The Hitchhiker's Guide to the Galaxy* and have an extra head grafted on. A bit ungainly, but at least you're never short of stimulating conversation. In Chapter 8 of this book we'll suggest another reason why you might want to contemplate a whole body transplant.

robot arm through thought alone. Scientists at Duke University Medical Centre in North Carolina implanted a series of electrodes into the part of the monkey's brain that they knew controlled movement (the motor cortex). By recording the output of these electrodes, they plotted the brain's electrical activity when the monkey performed specific tasks, such as reaching for small bits of food.

They analysed this data on a computer and worked out which signals corresponded to which movements. Every specific movement would produce a particular pattern of electrical activity. Once they had got the analysis right they decided to hook the computer up to a robot arm. In a series of experiments, they managed to get the brain signals from the monkey, once processed by the computer, to move the robot arm. Not just up and down, but in three dimensions. It even worked over the internet from a site 600 miles away from the lab.

The design of the monkey's brain helped the scientists out. The experiment demonstrated that there was not one specific nerve cell, or neuron, for each tiny movement. Instead, clumps of neurons worked together to produce a particular behaviour, such as extending the arm. For each movement the monkey made, the electrical activity showed up across a relatively large area of the motor cortex. This means that the electrodes did not have to be connected to one particular neuron to get results. They did not even have to be in an exact position, just spread broadly across the correct part of the brain.

If we want to place silicon implants in our own brains, all this is quite encouraging. We do not have to know the function of each individual neuron in the brain, just the function of particular regions. There are, of course, some significant ethical problems with this sort of research, but we will quietly brush these under the carpet for the moment. If you're contemplating life as a Dalek then disdain for morality is probably all to the good.

CAN YOU FEEL THE FORCE?

A monkey that can control a robot arm is quite an impressive demonstration of how you can move a limb through thought alone but it's very much a one-way process. The monkey is not getting any feeling back from the robot arm.

Likewise, Kevin Warwick could control a robotic hand and use it to grasp objects, but he couldn't *feel* those objects.

In humans the two processes of control and feeling go together. If, for example, you reach out and pick up an egg, in order to make the series of precise movements required it is important that you feel what you're doing. If you can't feel the egg and tell that you're clasping it using just the right amount of force, then you'll most likely end up holding a gooey mess. While one set of nerves is operating the hand, another is providing sensory information back to the brain. This process of feedback is crucial to successful cybernetics and is a whole field of study in itself, known as haptics, from the Greek for 'touch'.

Without some sort of feedback mechanism a body upgrade is going to prove socially disastrous. Imagine the embarrassment at business meetings when you shake hands and remove a couple of your client's fingers. Trips to the toilet for men would turn into the worst kind of Russian roulette. And any sort of intimate relationship is likely to quickly come unstuck. Literally. It's not just the practicalities of movement that make touch desirable. Without feeling, we would become isolated from the world around us – no direct contact at all with other human beings, from kissing to kicking and everything in between (and beyond).

Simulating the touch of a human hand is anything but simple. Just the tip of each of our little fingers contains around 2,000 receptors. These can detect shape, texture and friction. Alongside are sensors for heat and cold. Each fingertip is continuously sending at least 1,000 nerve impulses back to the brain every second. Multiply that across the whole hand and you begin to see the challenges involved in trying to replicate sensation electronically. It is even more impressive that the brain can decode all these tens of thousands of signals and make adjustments to movement accordingly. Sometimes we are aware of what we are touching, sometimes not. We can pick things up without consciously knowing how.

Remote-controlled robots are often used instead of humans inside toxic environments, such as the radioactive areas of nuclear power plants. Devices

Virtual cyborg

Fed up with the reality of life? Frustrated that your cyborg dream is floundering? Don't have $6 million? Fear not, you can escape to your own *virtual* reality. In the film *The Matrix*, machines take over the world and enslave humans to use them as biological batteries. The reality is bleak but for the humans in the matrix the virtual reality is perfectly normal – it looks just like the world we are all living in now.

This raises the question how do we know we are not living in some flawed artifice devised by our digital masters? Dr Nick Bostrom, a Swedish philosopher at Yale University, reckons there is a one in four possibility that we are already living in a matrix. He reasons that if civilisation were to develop to the stage where it could develop artificial consciousness, the artificial consciousness would need to inhabit a universe. This might be it. Alternatively, we could all be self-aware computer simulations in some intergalactic games console.

If that makes your head hurt, how about this. If we assume we are all inside a vast virtual reality anyway, then how about we escape from it with our own virtual reality. Science fiction is full of virtual situations, the Holodecks of later series of *Star Trek* being the perfect example. Star fleet officers act out their every fantasy in a large empty room transformed into a virtual world. It turns out their fantasies are fairly mundane. We suspect that if you were really stuck on a spaceship with a load of holier-than-thou types in polyester uniforms, you would opt for something more exotic. Or take drugs.

To create a virtual environment, your brain needs enough stimulation to convince it that what it's experiencing is real. This means some sort of haptic suit so you can feel the virtual environment, glasses that display the virtual environment and suitable devices to provide sounds and smells.

The hardware is already available. Haptic devices are increasingly being used in arcade games, and the home computer is likely to be next. Just imagine, you'll be able to feel the size of documents, touch images on websites and experience the full horror in spectacular 3-D surround sound when your machine crashes. That's progress.

like these need to employ haptics to give feedback to their human operators so they can accurately manipulate objects from a safe distance. One of the most advanced haptic devices is the CyberGlove, used by scientists at the Stanford Dextrous Manipulation Laboratory (yes, really) in the United States to control a two-fingered robotic arm. Rather than 1,000 sensors, the robot has only a few electronic ones, but it does nevertheless feed back force to the operator. They can feel what the robot feels.

Other scientists, notably those at the MIT's 'touch lab' are looking at ways of using haptics in surgery. Remote-control devices are already used in operating theatres, particularly in keyhole procedures, in which the incision in the patient is very small. Surgeons can manipulate instruments remotely using video monitors to see what they're doing. How much better, the researchers argue, to use machine haptics so the surgeon can *feel* what they are doing as well.

ROBO RATS

So if we want to connect a cybernetic device directly to our nervous system and ultimately our brain, we don't just need to be able to control it, we need to be able to *feel* it. This is why researchers in New York have developed remote-controlled rats. Really.

The rodents sport little backpacks equipped with a miniature TV camera, batteries, a transmitter and little T-shirts reading 'Have a nice day' (actually we made that last bit up). The operators stand a few hundred metres away, controlling the robo rats from a computer keyboard. By tapping the right keys they can send the rats in the desired direction. It's a system based on rewards and punishments, but rather than using lumps of cheese, the scientists at the State University of New York Downstate Medical Center are using direct connections to the rats' brains.

The researchers implanted electrodes in the areas of the brain that control whisker sensations. When these areas are stimulated, the rats feel a touch on a particular whisker. If the controllers want the animal to move right, they use the electrodes to 'virtually' touch a right whisker. If the rat obeys the instruction, it will get rewarded with another stimulation, this time to the pleasure centre of its brain. After a period of training, the rat will move in the desired direction because then it gets rewarded.

The idea is to use these rats either to investigate collapsed buildings for survivors or as rodent spies, able to penetrate areas inhospitable to humans. The technology involved is relatively simple and the rats relatively expendable. After training, they usually go where the researchers want them to, because it's pleasurable.

The goal of the research is not rat enslavement but to investigate how touch can be stimulated in the brain. The fact that it is proving successful bodes well for any attempts to feed back feelings from artificial limbs into human brains. Of course, before you do that, you have to know where to stick the electrodes.

ALL IN THE MIND

A hundred and fifty years ago the brain seemed remarkably simple. By the beginning of the nineteenth century, scientists were more or less agreed that it was the brain, rather than the heart or liver, that controlled the rest of the body. The brain deciphered our senses, determined our mood and held our memories. It led us to love and hate, cherish and murder.

A Viennese physician, Franz-Joseph Gall, had even worked out which bits of the brain did what. His theories, which were later to become phrenology, held that the brain was divided into 26 zones that he called 'organs'. Each of these was specific to a particular emotion or function. There was one for cautiousness, one for friendship and another for self-esteem and so on. If you touch your skull about four centimetres above the left eye, that's where you'll find 'Gall's organ of mirthfulness'. Murderers even had an organ that pre-ordained them to kill people. You can still buy the ceramic heads today, mapped out with the various phrenological organs. And although phrenology is now considered very much a pseudoscience (i.e. rubbish, like astrology), it turns out that there *are* certain parts of the brain that do specific things.

Reading the signs

Phrenologists claimed that the size of the brain's 26 organs determined the shape of the skull: the bigger the bump, the larger a particular trait. This meant that experts could tell someone's personality, intelligence and predisposition to murder by feeling their head. Remarkably, the 'science' of phrenology persisted well into the twentieth century and was particularly popular during the American depression thanks in part to a wonderful invention, the psychograph.

Created by American Henry Lavery, the psychograph resembled one of those hairdryer helmets that women in curlers used to sit under at the salon. It was attached to a wooden cabinet, inside which the result was printed. The person being 'read' would sit under the helmet and 32 small probes would be lowered onto their head. These touched the various areas of the skull corresponding to different mental faculties. Each probe was attached to five electrical contact points in the headpiece, the shape of the head determining which contacts were activated. Once everything was ready, the operator would pull a lever, an electric motor would whirr and the reading would begin.

Printed statements listing 32 attributes were mechanically stamped with a score depending on the position of the probes against the subject's head. Each attribute was on a scale of one to five, one meaning 'deficient', five 'very superior'. The machines were popular in the foyers of cinemas and department stores and initially made the inventor considerable amounts of money. To begin with, people believed in what they were being told. In the end though, scepticism finally won through.

MEMORIES ARE MADE OF THIS

Some parts of the brain deal with the senses, such as sound and sight. Other areas process speech or control movement of limbs. Tucked underneath the main bulk of the brain is the limbic system, which generates our emotions and controls our urges and appetites. Scientists have found that stimulating particular areas of the brain can lead to particular behaviours or feelings. (We'll talk more about the brain in Chapter 8.) One component that has been particularly well studied is the hippocampus, which is where long-term memories are formed. It's here that scientists believe they can make a direct electronic connection to the human brain.

The hippocampus is part of what is sometimes called the unconscious brain, as we are not aware of anything it's doing. Close to the junction with the spinal cord, the hippocampus seems to turn our experiences into memories, it *encodes* them and sends them off to another part of the brain for long-term storage. It is not something that should be taken for granted. Sufferers of strokes, Alzheimer's and epilepsy often lose the ability to form new memories and until now there's been nothing much medical science could do about it.

After more than ten years' research, scientists at the University of Southern California have come up with an alternative: an artificial hippocampus. They have yet to test it, but if successful it will be the first part of the brain that can be replaced by electronic technology. The scientists, led by Dr Theodore Berger, were somewhat hampered by the fact that no one knows how the hippocampus works. Like much of the brain, it is a bit of a mystery. So instead of attempting to figure it out, they decided to copy its behaviour. Like the neurons and neural nets we talked about when building a domestic goddess, the hippocampus has electrical inputs and electrical outputs. By feeding in known electrical inputs, the scientists figured they could measure the corresponding electrical outputs.

Bad news once again, we're afraid, for rat lovers. The researchers took tiny slices of hippocampus from the brains of rats. By sticking electrical signals through them, they were eventually able to work out which particular electrical inputs produced which outputs. Using this information from all the slices, they constructed a computer simulation of how the hippocampus behaves. They could then programme this information into a computer chip. Although the chip will almost certainly work in a different way, it should still be able to replace the hippocampus, receiving signals from one part of the brain and encoding them to be stored as memories in another.

The scientists are now in the process of testing the device, first on rats, then on monkeys. They plan to remove part of the animal's hippocampus and replace it with the electronic prosthetic. If you're uneasy about animal experiments, the field of cybernetic implants probably isn't for you. If all goes well, the first human transplant will take place in a few years' time. For someone who can no longer remember new things, the artificial hippocampus could be life-changing.

But what if the prosthetic worked too well? What if it let you remember the completely useless stuff, like all the words of every advert you saw on TV last night? Worse still, if it gave you a clear memory of things you'd rather forget – traumatic events like death, divorce or failure to qualify for the European championship.

Although our ability to forget things can be extremely annoying, it's also a way of keeping us sane. If we kept in our minds the details of every bad thing that had ever happened, our lives would be seriously impaired. Perhaps fatally.

The artificial hippocampus experiments demonstrate the potential of a direct connection between the brain and electronics. Using the same principles, it might be possible to develop a device that didn't just replace a part of the brain, but enhanced it. A more advanced artificial hippocampus holds the potential to help us learn new things by downloading information directly from a computer into our minds. The right software could transfer whole documents, new languages or skills straight into our memories. It's an idea used in the *Matrix* films, in which computers impart knowledge of everything, from martial arts to the piloting of a helicopter, directly into the human brain.

Putting a chip in the brain seems so obvious it's a wonder everyone isn't doing it. There has to be a drawback.

SELF-DESTRUCTION

If you remain convinced about the benefits of cyborg technology, are not too concerned about the ethics and are ready to sign up for some gory surgery, you should be aware that it could go horribly wrong. The body will almost certainly fight back.

The body's immune system will attack any 'foreign' object, from viruses, to bacteria to artificial implants. Quite apart from the risk of infection, any experiments such as the one conducted on Kevin Warwick or the numerous rats and

Cyber vision

Every cyborg worth its salt has better vision than its human equivalent. The bionic man could spot a scantily clad distressed female across vast distances with his bionic eyes. The Borg in *Star Trek* also have perfect vision. What if the same type of technology that produces an artificial hippocampus could also be used to upgrade eyesight by routing a digital camera through a chip in the brain?

Not only would a camera be able to see over much longer distances than the human eye but it could enable you to switch to different types of vision, giving you infrared sight in the dark or even X-ray vision to really freak people out. Incidentally, if you did have X-ray vision you wouldn't be able to see through people's clothes to their underwear, just their bones. And the bones of a very attractive person don't look much different from those of an ugly person. Unless you are an orthopaedic surgeon of course.

In the first instance, scientists are trying to restore vision to those who have lost their sight. Millions of people go blind each year because of diseases that damage the retina. This is the area at the back of the eye where images form, and it contains cells called photoreceptors.

There are two types of photoreceptor, rods and cones, both of which convert light hitting the retina into electrical nerve impulses. These are funnelled into the optic nerve and travel to the brain, where images are formed. (What the brain actually 'sees' is slightly more complicated, but we'll come to that in Chapter 8.)

Teams of scientists around the world are looking at ways of replacing damaged photoreceptors. No one can yet claim to have restored sight to the blind, but some experiments are producing promising results. An American company, Optobionics, has been carrying out clinical trials on an 'artificial silicon retina'. Only 2 millimetres across, it is implanted directly onto damaged photoreceptors. The chip is covered in thousands of microscopic solar cells, which, like the biological cells they replace, convert light into electrical impulses. The few patients who have received the implant all report improvements in their vision. One man, for instance, was able to see his wife's face once more.

Scientists at the American Space Agency, NASA, are taking a different approach. Instead of solar cells embedded in silicon, they are looking at using thin ceramic films that are sensitive to light. The entire detector is made up of 100,000 individual light-sensitive cells, each almost exactly the same size as the rods and cones they are replacing.

There is a long way to go before either of these technologies becomes widespread. All the methods under development rely on the nerve cells behind the retina remaining intact. Unfortunately, scientists have little idea how our brain interprets the signals from the optic nerve. We're afraid that hooking up a video camera directly to the brain is not yet possible.

monkeys we've talked about will almost certainly come up against the immune system. Usually this means the chip will stop working after a few days or weeks as it becomes coated in immune cells. By the end of Warwick's experiment, only one out of the original 100 electrodes was still functioning. An implanted silicon chip is treated no differently from a pathogen.

Twenty-four hours a day there's a war going on inside our bodies. Usually we've got the upper hand and the body overcomes bacteria or viruses quickly, so we don't even notice. Sometimes the attackers win a temporary victory and we come down with a cold or flu. Occasionally the body can't cope and we need drugs such as antibiotics to help us out. Finally, there are viruses such as HIV, which directly target the immune system. Then we're really in trouble.

The human immune system has evolved an incredible complexity. It even has its own distribution network, centred on the lymphatic vessels, a special circulatory system carrying a clear liquid called lymph (Greek, incidentally, for 'pure, clear stream'). This network is dotted with lymph nodes, which act as meeting places for the various immune system cells.

There are a whole load of different types of cells and several different lines of defence the body utilises to repel invaders. The components are manufactured in the bone marrow, found in the long flat bones of the body, such as the pelvis. Key to the whole process are cells called lymphocytes, which come in two different forms, B and T.

The B cells produce protein fragments that bind to the surface coating of invaders, or foreign antigens as they are known. These proteins either disable the antigens or mark them for destruction. What's remarkable is that these small pieces of protein bind with very specific proteins on the surface of the antigens. The immune system has to constantly evolve to cope with new threats.

The T-cells are the killers. They can spot the enemy according to the types of protein on the cell surface. The body's own cells are not attacked but others (such as those marked for destruction by the B-cells) are. The targets are ambushed and destroyed. It's a cruel world in there.

To the immune system, the hard surface of an electrode protruding from a silicon chip is most obviously alien and is marked for destruction. Researchers have estimated that an electrode is around a million times harder than the tissues in the brain. While the immune system cannot destroy the implant, it does end up impairing and eventually halting its function, covering it with layers of immune cells. The crude way around this is to hold off the body's defences using drugs called immunosuppressants. Unfortunately, this has the effect of lowering the body's defences against everything else, like infections. And when it comes to silicon chips it's also not particularly effective. The answer is to make electrodes more like the tissues of the body. Engineers at the University of Michigan have started to develop electrodes that are soft, strong and fluffy (being very long would possibly also be an advantage).

To overcome the hardness problem, the material could not be any type of metal but still has to be able to conduct electricity. Researchers led by David Martin started experimenting with plastics known as conducting polymers, which are long chains of molecules made up mostly of carbon. They found they could get these to cling to the electrodes, providing a bridge between the hard surface of the metal and the soft, wet surface of the brain. To make the electrodes fluffy they used a material called hydrogel, which strongly attracts water molecules. Under the microscope, the electrodes appear to be covered in fur. A process of freeze-drying the mixture of hydrogel and polymer onto the electrode makes for a strong connection. Once implanted, the gel re-inflates to become soft again, and the neurons can grow over it.

This process of developing 'fuzzy polymers' is at an early stage and scientists are only just getting round to the rats. But work like this is essential if cyborgs are ever to become a reality.

MUSCLE

It's all very well using a chip to enhance mental capacity but if we want to use our cyborg powers to make new friends and influence people or take over the galaxy (delete as applicable) what we really need is brawn – some enhanced physical powers to match our mental ones. That is, assuming you haven't spent all the allocated $6 million yet.

Unlike many of the previous human attributes we have talked about, such as hearing, vision or touch, when it comes to brute force, machines are undoubtedly our superior. Motors and hydraulics can shift and carry a great deal more than even the strongest human. Often, though, with mechanical technology, strength is related to size. A crane that can lift 40 tonnes is considerably bigger than a crane that can only lift four tonnes. It needs to have a stronger frame, thicker cables and a more powerful motor. The same is true when applying that sort of mechanical technology to humans. To become superhuman, everything needs to be rather larger than can be conformably accommodated within the relatively narrow confines of the human body.

Designers of real and imaginary robots have also had to grapple with this problem. Look at the glistening golden protocol droid of *Star Wars*, C3PO. All his joints are mechanical, with little rods passing from the outside of the upper arm to the surface of the lower arm. These act as levers, raising and lowering the limb. If we used levers in this way, we too would have rods passing across our elbows.

The other problem with motors and hydraulics is that they use considerable amounts of power. The bigger the motor, the larger the batteries. If you are planning to take over the galaxy, it's a bit embarrassing if you have to keep excusing yourself to plug into the nearest power socket. For all these reasons cybernetics researchers are once again looking to the human body for inspiration.

As you read this book hundreds of muscles are contracting. Heart muscles are pumping your blood around; your gut is pushing your last meal through the system and tiny muscles attached to your eyeballs are stretching and contracting as you follow this sentence. Even the hairs on your arms are each controlled by individual muscles. The whole human skeleton is covered with muscles, which makes our movements incredibly subtle and flexible.

For obvious reasons, we have no direct conscious control over heart muscle or the muscle lining our digestive system. But we can direct our skeletal muscle to move our limbs or bend our body. Skeletal muscle is made up of thousands of individual fibres, criss-crossed by blood vessels and nerves. The number of fibres each person possesses seems to be fixed early on in life, it's just the thickness of the fibres and the other tissue around it that can increase. So you might have the same number of muscle fibres as the fittest Olympic athlete (the key word here is 'fittest').

Unlike nerve fibres, muscle fibres are not made up of individual cells but rather a fusion of cells. Under a microscope, each fibre is divided up into a series of identical blocks. Each one looks like a pair of interlocking combs. If we stick with the analogy, then the teeth of one comb consist of filaments of a protein called actin, the teeth of the other filaments of a protein called myosin.

When muscles contract, hooks on the myosin filaments grab onto the actin filaments and pull forward, so they slide past. This process is repeated, each time reaching further forward and pulling. Imagine hauling yourself up a wall, reaching out with your hands all the time so you pull yourself up. It's a bit like that. With millions of actin and myosin filaments all working together, the entire fibre contracts and the limb moves.

Copying such an intricate system in order to develop artificial muscles is quite a challenge. So rather than try to build an exact replica, scientists have been looking at materials that behave in the same way: substances that can be

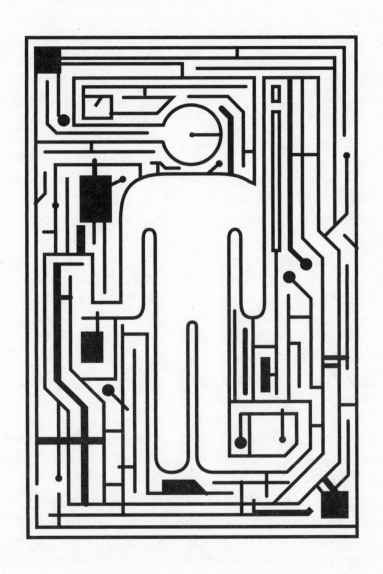

Batteries not included

Surf the web using a search engine for the film *The Matrix* and you'll get almost 1 million results. This not only says a great deal about the web, but a considerable amount about the culture surrounding the films. One of the hottest topics of discussion surrounds the premise that we could already be living in a virtual world. Another concerns the science. The storyline explains that humans are only being kept alive as biological batteries feeding a master race of machines.

As a concept it gets quite a few scientists most upset. To make the whole process worthwhile would mean getting out more energy than is put in. This breaks one of the most fundamental physical laws, the Second Law of Thermodynamics, the gist of which is that you can't just conjure up energy from nowhere. Why are we telling you all this? Because if you intend to become a cyborg, you are going to need a lot of extra power.

Artificial limbs, cochlea implants and smart clothes all need to be hooked up to a reliable supply of electricity. That invariably means connecting them to a battery belt, which can often be bulky and cumbersome. Fine if you're a Dalek, not so good if you're planning a future as a lithe superhero. If you're not too fussed about side-effects and can lay your hands on the raw materials, nuclear power might be one way around the problem. Space probes like the Saturn craft Cassini use radioactive isotopes which generate heat to produce electricity. Although it's one thing to send them to the other end of the solar system, it's quite another to wear them to the shops.

Until batteries get a lot smaller, any cyborg parts need to be designed to consume only a small amount of power, which rules out a lot of the more mechanical technology. Small silicon chips in our brains have only minimal electrical requirements, as they are integrated into the nervous system. Which brings us back to *The Matrix*. Ideally, we want to use the energy from our own bodies. Energy we get from food.

We don't burn food to make energy. After food has been digested, a process of energy generation takes place in a component (organelle) of every single cell. This component is called the mitochondria. Energy enters cells as molecules, chemicals broken down from food or the body's food stores (such as fat). Each molecule contains individual atoms connected by chemical bonds. In the mitochondria these bonds are broken, a process that releases energy. This energy is then immediately used to manufacture another molecule called ATP.

ATP, or adenosine triphosphate, is the body's equivalent of a battery. It can be carried around the cell and easily broken down to produce energy where it's needed. Trying to create a large device that runs off ATP is difficult. One way is to employ the techniques of nanotechnology and incorporate electronics into individual cells or cells into silicon chips. Researchers at the University of California have already tried this latter idea and are working on ways of controlling cells using digital technology.

As to *The Matrix,* in which humans are used as batteries, the controlling machines would be better off burning the food they're feeding the humans. Alternatively, if they're just after an organic form of electricity they could stick a couple of electrodes in a lemon.

stimulated to contract. One of the most promising is a type of plastic known as an electroactive polymer. Like a human muscle, it bends and flexes when an electric current is applied to it. But it works in a completely different way.

The polymer looks like a thin ribbon. It is made up of repeating chains of molecules, similar in structure to the conducting polymers we have already talked about. When it is given a burst of electricity, charged particles on one side are pulled together, making that side shorter. Particles on the other side are pushed apart, making that side longer. The overall effect is a bending of the material. Electroactive polymers do not need nearly as much power as electric motors and can be kept small. They are already being used to develop small robots which NASA hopes could be useful for robotic landing craft on space missions, where power is at a premium.

Other scientists reckon they've got an even better material, based on a molecule called a carbon nanotube. A few years ago, if science pundits were to be believed, carbon nanotubes would soon be revolutionising our lives. 'Soon'

The real fantastic voyage

The sci-fi kitsch classic *Fantastic Voyage* is a movie in which a miniaturised submarine and its crew journey through a body to try to save it – battling antibodies and a turgid script along the way. Although the film itself does not age particularly well, the idea of using tiny machines to scramble around inside us has some merit. So much so that scientists are already devising ways to make it possible.

This area of research is known as nanotechnology, named because it is a technology and is very, very small (anything 'nano' is a billionth). Rather than connect up a chip to a particular area of the brain, nanotechnology has the potential to integrate man-made components into every single cell in the body. A team of nano-bots could patrol your bloodstream seeking out trouble or repairing damage. The Borg from *Star Trek* are full of them, as every one of their cells is specially adapted.

Nano-machines are constructed using the basic building blocks of stuff: atoms and molecules. At Cornell University a team led by Carlo Montemagno has built a motor five times smaller than a red blood cell. The researchers took a protein from an E-coli bacterium and a piece of nickel and attached both to a propeller 100,000 times thinner than a human hair. What's even more remarkable is that it uses the same power source as a cell: ATP, meaning that it can be fully integrated into the body.

The immediate aim of much of the research is for the ongoing treatment of health disorders. One implant, developed by scientists at the University of Illinois, can automatically deliver insulin into the bloodstream of diabetics. The device contains cells from the pancreas of a pig, which, like cells in our own pancreas, produce insulin if they detect too much glucose in the blood. The implant has been engineered so glucose molecules can enter but antibodies can't. That way the body's defences won't attack it as alien.

There have been some great promises made about molecular nanotechnology, reflected in the millions of dollars invested in its study. Biological cells are little more than collections of molecular machines, the argument goes, so why can't we design new ones? One of the main exponents of this vision is a scientist called Eric Drexler. He's talked of machines that can travel around

> the body repairing damaged cells or correcting errors in our genetic material, DNA. Finally, perhaps, a cure for cancer.
>
> Ultimately, the visionaries expect nano-machines to reproduce, becoming a life form in their own right, co-existing with the human body. Of course, once you release a swarm of self-replicating machines into your bloodstream there is no going back. With simple programs to control them, a more complex behaviour is likely to emerge when they start working together. And what happens then is difficult, if not impossible, to predict. In his latest book, *Prey*, Michael Crichton develops just such a scenario, in which self-replicating nanobots go out of control, becoming human predators.

hasn't happened yet; nevertheless, they remain extremely interesting molecules. They consist of phenomenally small cylinders of carbon atoms, around a millionth of a millimetre in diameter. When a load are stuck together the whole structure is both strong and flexible.

Researchers at the Max Planck Institute in Germany found that when they applied an electrical charge the tiny tubes expanded; when it was removed the tubes contracted. Just like muscles. The experiment was conducted on a sheet of nanotube 'paper' attached to a piece of sticky-back plastic. Which just goes to prove that we should all have paid more attention to *Blue Peter* ('Today Goldie and I will be clearing leaves in the sunken garden while Lesley explains quantum theory using an empty washing-up liquid bottle and some tin foil.') It also demonstrates that they are a long way off producing an artificial muscle. In the meantime technology like this could be used in reverse to generate electricity. The flexing of carbon nanotubes might be a good way of turning the power of ocean waves into electric currents.

So what of our cyborg ambitions? Technology involving polymers or nanotubes would seem to be a much better way of moving cybernetic limbs than motors and hydraulics. But the science is relatively young. If you've already got a full complement of working arms and legs, it's probably best to hold on to them for the moment.

BUILDING A BORG

The first thing anyone looking to design a cyborg from scratch will realise is that the human body is a wonderful machine. So far, attempts to replicate its parts have produced inferior results. Even so, such replicas can make a huge difference. Efforts to help the deaf hear using cochlea implants, or the blind see by devising artificial retinas will almost certainly improve people's lives. And artificial limbs are getting more and more like the real thing all the time.

At the moment, rather than being better, stronger, faster, any bionic man would be worse, weaker, slower. Which not only scans poorly but isn't much of a selling point. Certainly not a good investment for $6 million. The good news is that, unlike in some other areas of science we have talked about, there are no fundamental problems to overcome to realise the cyborg dream.

For an immediate upgrade it is probably best to stick to technologies that utilise the senses you are born with. Smart glasses could enhance your view of the world while your pants keep the room around you at a comfortable temperature. A PDA or PC incorporated into a jacket could alert you to meetings and world news while on the move. This has the double effect of making you seem incredibly busy and important while your pants keep you cool under pressure.

In the near future, a chip implanted directly into the brain will certainly be possible, and perhaps desirable. Then things would really start to get interesting. How would it feel to be directly connected to the rest of the world via the web, mentally surfing a network of the world's knowledge, information and misinformation ...? Communicating with similarly connected people through thought alone ...? Remembering everything and forgetting nothing ...?

If enough people are interconnected, perhaps a swarm-like consciousness will emerge between different people right across the world. Like the Borg. Once our minds are connected will we ever want to, or indeed be able to, disconnect ourselves? It could be argued that we would no longer need our bodies. We would be living a virtual life in a higher plane of existence, abandoning the material world altogether.

Several cybernetics guru types have thought long and hard about these ideas and their almost limitless implications. One, Hans Moravec, a pioneer of robotics, talks of humanity merging with machines to create new, superior forms of life. He imagines them constantly rebuilding themselves and ultimately becoming pure mind with no physical presence at all. A cyborg utopia.

If these types of scenario are born out, it's likely that only a certain proportion of humanity will be able or want to become more than human. A counterculture of 'norms' would probably emerge to challenge the cyborgs. Would cyborgs triumph over humans, as Homo sapiens did over the Neanderthals? Or would it take just a simple computer virus to wipe them all out for ever?

Of course none of this might happen. There are two, not unimportant, barriers to cybernetic technology.

The first is ethical. Certainly in the UK, many of the animal experiments we have talked about would never be permitted. Try going to a funding council with a proposal to build a cyborg and they will say you are crazy. There are also issues over who controls the technology. For instance, the artificial hippocampus that we talked about could benefit people who have lost the ability to form new memories but these are exactly the people who would be unfit and unable to give permission for the transplant.

The other problem is psychological. Cyborgs look strange. Most people find it very difficult not to stare when they see a person with an arm or leg missing or even someone sporting an artificial limb. How would we as a species adjust to having electronically enhanced physical features? There's nothing pretty about the Borg (unless you're into that sort of thing). Would we take to having wires sticking out of our heads or is the whole idea repulsive?

What's certain is a definite trend towards integrating humans with machines. It's possible that the increased user-friendliness of computers will extend to a direct link with our electronic counterparts. If so, this will have a profound impact on society. As for using technology to improve our physical capabilities, it's already out there in the form of cochlea implants, pacemakers and artificial retinas. As for improving on what we've already got – if it ain't broken, don't fix it. In the meantime you could always get a sink plunger and pretend.

CHAPTER SEVEN

How to Remove an Eyesore

An eyesore means different things to different people. It can be a piece of art (remember that unmade bed?). It can be a building (wander along London's South Bank and take your pick) or even a whole town.

'Slough' was written by John Betjeman in 1937 after the town had gone through a period of industrialisation. Betjeman wasn't too happy with the result as the first line of the poem memorably asks for some 'friendly' bombs to fall on the place. The late poet laureate obviously decided that no amount of rebuilding could remedy the situation. Some of us, no doubt, have felt the same about other areas of Britain at some time or other – the Bull Ring in Birmingham, for instance, or indeed Birmingham itself.

Today Slough has much more to offer. It is now the home of Europe's largest trading estate while the Bull Ring is undergoing a long overdue makeover. But think how much easier it would be if all other huge eyesores – your version of Betjeman's Slough – were simply removed from view. Just like that. After all, when so much effort has gone into producing that new slimline upgraded body, your surroundings need to make the grade too. This is the chapter that will tell you how you can bring this about, by making things you don't like disappear. All it requires is the careful and extremely controlled use of a black hole.

COSMIC WHIRLPOOLS

A black hole is an invisible cosmic whirlpool. Get too close and you will be gradually pulled towards it, swirling around in ever decreasing spirals accelerating around its central point. Viewed from afar, it would be similar to watching bath water drain down a plughole except that this whirlpool uses gravity not water. And unlike a dropped ring, which, with a bit of plumbing, can be retrieved from a plughole, nothing can escape from the centre of a black hole.

Let us assume you ignored the 'Beware of the black hole' warnings and did get too close. You are now spiralling towards the black hole feet first. As if things weren't bad enough, at some stage in your dizzy journey there is an invisible boundary with a sign saying 'Point of No Return. Abandon Hope All Ye Who Enter Here' and 'No Junk Mail Please'. Once past the sign your friends, who are watching from a spacecraft at a safe distance, see strange things happen. You will appear to stand still, frozen in time, before simply fading from view. Sadly, this is not how you see things as you attempt to fire a distress flare. Instead, you keep plunging deeper into the black hole until its gravitational forces are so strong that your body is squeezed and stretched to infinity. No one will hear you scream. No one will witness your gruesome death by what some scientists have called 'spaghettification'. And no one will ever see the light from that distress flare.

Scared? You should be. The reason for this miniature horror story is to get one thing straight. Black holes need to be handled with care. Consider them as the astronomical equivalent of circus lions. One false move and you'll lose a lot more than a chair leg. If we are going to use a black hole to remove an eyesore, we've got to make sure it's done properly. Otherwise the black hole could start to eat its way through the Earth and take out a lot more than Slough. So let's give black holes the respect they deserve and gradually get to know these Darth Vaders of the universe.

Surprisingly, considering that they've become one of the mainstays of twentieth-century science fiction, the concept of black holes originated more than 200 years ago. At the end of the 1700s, John Michell in England and Pierre Simone de Laplace in France independently used Newton's laws to work out the escape velocity of a specific object. Escape velocity is the speed something needs in order to escape the gravity of a planet or star. It depends on the planet or star's mass and size and is the same for any object trying to escape – be it a rocket or a cricket ball. For the Earth it is at least 11 kilometres (7 miles) per second. Any rocket must therefore reach or exceed that speed in order to propel itself into orbit. The escape velocity for the moon is significantly lower, at only 2.4 kilometres per second, so any spacecraft sent from the lunar surface would require a lot less fuel than if it was launched from Earth.

Michell and Laplace both decided to examine an imaginary object: a star whose matter was compressed into an extremely small space, making it very dense. They worked out that if it was small or compressed enough, its escape velocity would need to be faster than the speed of light, which would mean nothing could escape from it, not even light itself. They suggested that objects like this would resemble an 'invisible star'. Centuries passed before scientists realised that these men in breeches had described something like a black hole. Not surprisingly, perhaps, given that physicist John Wheeler didn't invent the astronomical term 'black hole' until the late 1960s.

These theoretical compact objects, or 'invisible stars', eventually reappeared when Albert Einstein published his General Theory of Relativity (also known as his theory of gravity) in 1916. The same predictions resulted: a very small and dense object would have such a strong gravitational field at its surface that nothing could escape from it, not even light. This is because the object's escape velocity would have to exceed the speed of light, but as one of Einstein's main tenets is that nothing can go faster than the speed of light, this would be impossible. A black hole is therefore like one of those insect traps or 'roach motels' sold in the United States. Or even the Bates Motel in *Psycho*. Because you can check in but you can't check out.

BLACK HOLES HAVE NO HAIR

Einstein's General Relativity, as we saw in Chapter 5, introduced a number of new concepts. The associated mathematics, known collectively as Einstein's field equations, had a range of theoretical answers, and scientists around the world immediately started working some of them out. Within months of Einstein presenting his theory of gravity, the first solution arrived describing a black hole.

It was from a German soldier fighting on the Russian front during the First World War. The soldier, astronomer Karl Schwarzchild, obviously needed to take his mind off trench warfare and so solved the mathematics of Einstein's General Relativity for a tiny spherical mass.

Schwarzchild's calculations resulted in the mathematical description of a static or non-rotating black hole with a central point. This is unusual in that it does not have a surface and instead has a boundary beyond which nothing can escape the black hole's gravitational forces. The boundary is that point of no return for our black hole explorer described earlier, also more commonly known as the event horizon. In astronomy the event horizon is a spherical boundary surrounding the black hole and, in memory of the man who first described it, the radius of this event horizon is known as the Schwarzchild radius. Schwarzchild survived the war, by the way, but contracted an illness and died several months afterwards. Life sucks sometimes.

Einstein would not accept the suggestion that these outrageously dense objects could exist, and other scientists agreed. But in 1963 the mathematician Roy Kerr came up with another black hole solution. This described a rotating black hole that had a region outside the event horizon (the ergosphere) dragging space-time along with it as it rotated. Instead of a single point, or singularity, Kerr's spinning black hole had a central ring.

Kerr's version also allowed something rather unusual. The central spinning ring could connect two black holes in an Einstein-Rosen bridge (see page 165) and theoretically allow someone to travel through it into a parallel universe. Whoever said mathematics was dull has never really understood the possibilities mathematics can offer. Since all the stars and planets in our universe have been observed spinning to some extent, Kerr's rotating black holes are more plausible than Schwarzchild's non-rotating ones.

Rotating and non-rotating models are now known as Kerr and Schwarzchild black holes respectively. Both types of black hole are completely described by their mass, charge (believed to be small, either positive or negative) and, if it's a Kerr-type black hole, rate of rotation (angular momentum). This relatively simple description – from a scientist's viewpoint – led to John Wheeler proclaiming that 'a black hole has no hair'. This is because, apart from mass, charge and spin, a black hole doesn't have any other characteristic (or 'hair'), such as temperature of composition, to complicate matters and make its smooth simple description hairy. Odd choice of words, yes, but in black hole terms, bald really is beautiful.

DEATH STARS

Black holes are predicted mathematically but by their very nature cannot be seen. Assuming that they do exist in our universe (and scientists are pretty confident about this), where do they come from? No one knows for sure, but, after extensive study of stars in our own and other galaxies, astronomers have firm ideas about how black holes form.

The Sun, our nearest star, is a massive ball of churning gases. Like most stars, it is made up mostly of two parts hydrogen to one part helium. Only the small amounts of other components – such as carbon, oxygen, nitrogen and metals such as iron and magnesium – vary from star to star. While extremely special to us, the Sun is in fact an ordinary star known as a yellow dwarf. The colour of a star reflects its temperature. Blue stars are much hotter than red stars, for instance. Stars have different ages too – and it pays to have a small mass if you want a long life. The massive stars live fast and die young as they burn up their fuel at a much greater rate.

The Sun's fuel will not last for ever, and it will eventually die. For the Sun, astronomers predict a relatively peaceful death. It will first expand into a red giant and then cool into a smaller white dwarf star with an extremely dense mass. But not all stars go so quietly. Different stars have different deaths, depending on their size.

You are my sunshine

We know a lot about the Sun. Astronomers have worked out its age, mass, size and brightness. These facts are crucial in their understanding of other stars throughout the universe.

Like an onion, the Sun consists of layers. The outermost atmospheric layer is the corona. This is the only visible part of the Sun during a total eclipse, when the moon completely obscures the Sun. Total solar eclipses can been seen somewhere around the world every 18 months; the last one visible from within the UK was in August 1999. If, like us, you were attempting to view it from a beach in Cornwall, then due to cloud cover you will be none the wiser, but some lucky people may actually have seen a corona and know what we are talking about. Most of the Sun's X-rays come from the corona and its temperature is a blisteringly hot 1 million degrees C.

Moving inwards, the next atmospheric layer is called the chromosphere. This is the major source of sunburn here on Earth, as it produces ultraviolet radiation. The photosphere is the nearest the Sun gets to having a surface, although you wouldn't want to loiter, as its temperature, although a lot cooler than the corona, is still around 5,700 degrees C.

The first layer inside the Sun is a region known as the convection zone, where hot gases carry energy from the Sun's core upwards and outwards. Images produced by solar observatories such as the SOHO spacecraft show that this layer appears granulated because loops of gases are moving up and down in convection currents.

The radiation zone, in between the core and the convection zone, is again named for the way the energy is transported. The closer you get to the solar core, the hotter the temperature and the greater the pressure and density. The core itself is at around 15 million C. Once inside, you are in the Sun's nuclear powerhouse. Here, hydrogen atoms are fused by a series of chain reactions into helium. Each second about 600 million tons of hydrogen is converted into around 596 million tons of helium. If you are wondering what has happened to the missing 4 million, it has become energy. Part of this energy is responsible for life on Earth, as it reaches our planet in the form of sunlight.

A star is kept stable through a balance of forces. One of these forces is pressure produced by the extremely high temperatures from nuclear reactions in the core which push outwards. Another force, that of gravity, pulls everything inwards. Towards the end of a star's life, when all its fuel has gone, the forces that keep it together become unbalanced.

It's sort of like trying to maintain some degree of stomach control by pulling in those underused tummy muscles and wearing tight trousers. If you relax and your stomach is larger than it should be, then pow. The buttons on your trousers will ping off and parts of your stomach will escape into places best left unseen. A star is the same. Instead of buttons, its outer layers get ejected into space. But if a star is the equivalent of opera singer Luciano Pavarotti's stomach, say, there's a massive explosion called a supernova. When that takes place, those flying buttons could take your eyes out. The star's insides, however, then collapse under their own weight and become a small densely packed core. The same probably can't be said for Pavarotti – but who cares with a voice like that?

Let's stick with our big belly analogy for a moment. With smaller stomachs the trousers always win the battle between the forces and embarrassment is avoided. Even though you have breathed out and let yourself go, the force from those tight trousers will push your stomach back in. It's the same with a smaller star, except that the gravitational force causes the star to collapse in on itself, and it becomes a hungry black hole.

The Sun is the equivalent of a small stomach, astronomically speaking. After an initial period of expansion, its core will collapse into a dead star called a white dwarf, about one hundred times smaller than its current size. So much matter will then be concentrated into such a comparatively small area that one teaspoon would weigh around 5 tons. By this time the Sun will have used up all of its hydrogen and, as the name 'white dwarf' suggests, will appear white. This is due to its high temperature. The hotter a star, remember, the whiter and bluer it looks.

Astronomers have guidelines to determine whether a star is likely to become a black hole or not. Those with larger masses (around three times the Sun's mass or more) will burn off their fuel until all that is left is an iron core, the ashes of a star. These stars explode as supernovae and shed their outer layers. Because the stars are no longer producing any outward pressure, they lose the battle against gravity pushing inwards. Some will form neutron stars, as so many neutron parti-

Little green men

Neutron stars, the highly dense stars formed by a supernova, were first seen by a Cambridge University graduate student in 1967. Jocelyn Bell and her supervisor, Antony Hewish, had built a huge radio telescope in a field to detect the radio signals coming from quasars. These are quasi-stellar radio sources, faint objects in the night sky that look like stars and had only been identified a few years earlier. Some quasars emit radio waves, and after several weeks' observation, Bell discovered a strange signal – 'a bit of scruff' – among the masses of paperwork and realised it could not have come from a quasar. It was pulsing for 0.3 seconds every 1.337 seconds, something that had never been seen before. Previously detected signals were always continuous. In all, she identified four of these signals and found they each pulsated regularly and rapidly

After briefly referring to these objects as LGM (short for 'little green men') in case they were messages from aliens, Bell and Hewish called them pulsars, as they appeared to be pulsating radio stars. Hewish suggested that pulsars were either white dwarfs or neutron stars. When astronomers Thomas Gold and Franco Pacini turned their attention to these mysterious objects they came to the conclusion that pulsars were rotating neutron stars. These stars emit radiation from their magnetic poles, and so if they are facing the right direction and rotating, these signals are intensified and are seen to pulse in the same way that a beam of light appears to flash on and off from a revolving lighthouse.

Hundreds of pulsars have since been located, shining their radio beacons around the universe, and all of these pulsars are rotating neutron stars. Hewish received a Nobel Prize for discovering pulsars. Astonishingly, Jocelyn Bell Burnell (as she is now known) was overlooked, although she has always received public credit for her role in this achievement. Burnell went on to have a distinguished scientific career and is now a fellow of the prestigious Royal Society and a keen promoter of women in science.

cles will be produced that they will bind together through gravity. Neutron stars are usually about 10 kilometres across and have an escape velocity of around half the speed of light. Unfortunately, if we tried to enlist a neutron star to remove Slough, its surface gravity would probably destroy the entire Earth. Oops.

Not all massive stars become neutron stars once they've died. There are stars whose layers might not get completely blown into space during their explosive death throes. Instead, some of this material may fall back into the core. Astronomers have calculated that if the mass of the core is more than three times the mass of our Sun, then it will implode into a black hole.

All of the star's mass would be concentrated into a small volume, making it very dense and with a gravity so powerful that not even light could escape. As gravity results from warped space-time, then space-time too should be infinite. If this happens at the point of singularity – the centre of a black hole where all the mass is found – then all the known laws of physics will break down. The British physicist Roger Penrose said that a 'naked' singularity – one on its own – could never exist, as it would always be clothed in an event horizon. He called it cosmic censorship, because we are unable to witness the laws of physics cracking up. No one has yet proved conclusively that this singularity definitely exists and whether there is indeed a boundary between our universe of space-time and nothingness. But the smart money wouldn't bet against it.

One other important thing to realise about a black hole is that it is not, contrary to popular belief, a cosmic vacuum cleaner that sucks up everything in its path. When you ignored the signs and got too close to that black hole at the start of this chapter, we mentioned that your friends were (no doubt smugly) watching your mishap from a safe distance. This is because the gravity of a black hole, while extremely strong because of the large mass squeezed into such an infinitely small point, still decreases with distance, the same as it does for other normal objects, such as planets or stars. If you double your distance from a black hole, for instance, you will feel a quarter of its gravity. A spacecraft could therefore orbit a black hole quite safely provided those onboard do their sums right and don't get too close. You have been warned. Twice.

BLACK HOLE HUNTERS

Hopefully, by now, we have convinced you that black holes are exactly what we need to make the world a prettier place. Looking for something that can't be seen isn't easy, but even so astronomers are certain that black holes are dotted around the universe. Finding one is firstly a process of elimination. There is a lot an astronomer can tell about something in the night sky just by looking at it through a telescope, measuring its brightness (luminosity) and using a spectrometer to split the object's incoming light into its constituent parts – just as rain droplets do with light to form a rainbow. By working out the object's mass and looking at the sorts of gases and materials it contains, astronomers can tell whether it's likely to be a star or a planet. But what if the object is not emitting light? How do you know if this invisible star is a black hole? The answer is, you look for indirect evidence.

John Wheeler – the man who invented the term 'black hole' – likened identifying a black hole to watching men and women dancing in a dimly lit ballroom. If the men are wearing black tuxedos and the women are in white dresses then you can see only women dancing. Despite not being able to see the men, however, you can predict their existence by the way the women spin and whirl around a central axis.

It is the same with a black hole. By examining the movement of objects around a suspect area in space, astronomers can conclude whether a black hole is likely to be there or not. One of a black hole's tell-tale signatures is a high-speed swirling disc of dust and gas particles, known as an accretion disc, being pulled into the invisible object by its gravity. The closer the matter gets to the centre of the black hole, the faster and hotter these particles get due to their friction and speed. Theory then predicts that some of these particles will get so hot that just before they disappear beyond the event horizon they will emit optical light or even invisible high energy ultraviolet light or X-rays – another potential black hole signature. This radiation will be too strong for a neutron star or a white dwarf and it will also fade away the closer it gets to the invisible star.

Convincing signs, definitely, but scientists still need proof that this is happening. Fortunately, there are a number of orbiting spacecraft that can help obtain a mounting pile of indirect evidence. Each spacecraft provides different pieces of the black hole puzzle, and all scientists have to do is to fit them together to see what sort of picture it makes. The following are some of the main black hole hunters in space.

- ❀ **The Chandra X-ray Observatory Center** is a NASA mission named after the Indian-born American astrophysicist and Nobel Prize winner Subrahmanyan Chandrasekhar. It was launched in 1999 to look at hot spots in our universe. These are areas in which there are colliding galaxies, exploding stars and black holes where high temperatures can be found. When dust and gas particles are swirling into a black hole they get so hot that they emit invisible high-energy X-rays just before they pass the event horizon and disappear for ever. As Chandra is the world's most powerful X-ray telescope, this makes it an ideal black hole hunter.

- ❀ **Integral** is the European Space Agency's (ESA) International Gamma-ray Astrophysics Laboratory. It went into service in 2002 and orbits the Earth once a day at high altitude – 40,000 kilometres and above – looking at high-energy radiation from space, specifically gamma rays. These gamma rays are mainly released during violent events in our universe, such as explosive supernovas – when matter collapses in on itself and, in certain cases, forms a stellar black hole. They are not easy to detect but Integral is studying the massive black holes in the centres of galaxies, including our own, for evidence.

- ❀ Other spacecraft observing X-rays include ESA's **XMM-Newton Satellite** (XMM for X-ray Multi-Mirror) and NASA's **Rossi X-ray Timing Explorer**. XMM-Newton's three onboard X-ray telescopes can make long observations of deep space while Rossi can detect minute changes in X-rays, over thousandths of a second. Together these two black hole hunters have helped to suggest that the X-rays coming from a black hole do come from the accretion disc of spiralling matter.

Hubble

Named after astronomer Edwin Hubble, the Hubble Space Telescope has produced some of the most beautiful and memorable images of our universe. It can see objects billions of times fainter than can be seen with the naked eye and has made enormous contributions to science for over a decade. Carried into orbit by a space shuttle, Hubble was designed in the 1970s and went into operation in 1990, on a joint mission between NASA and ESA. It is now considered an extraordinary success, despite the fact that it was publicly vilified shortly after its launch.

First of all, people questioned its cost – estimated at around $6 billion – and then, more importantly, its capability – after its primary mirror was found to be faulty. Thankfully, the mirror was eventually repaired and today few people doubt the benefits that Hubble has brought to our understanding of the universe. But in the same way that a computer is only as good as its software, Hubble is only as good as its equipment. For this reason it is constantly upgraded and regularly serviced by astronauts during shuttle missions.

Hubble is in a relatively low 600-kilometre (375-mile) orbit and circles the Earth above the distorting effects of atmosphere. Rather than a lens, the space telescope uses two mirrors: the largest is a 2.4-metre (7-foot) primary mirror to collect and focus starlight while a smaller, 30-centimetre (1-foot) secondary mirror reflects this light back to the primary mirror through a baffle, to eliminate any stray light, and into the onboard scientific instruments. Hubble's wings are solar panels which convert sunlight into the 2,400 watts of electricity needed to power the telescope.

Hubble is mainly used for observations in visible light, although it can also detect infrared and ultraviolet light. This is useful for black hole detection, as there is both visible and ultraviolet light in the outer part of an accretion disc. Its newest instrument, the Advanced Camera for Surveys (ACS), was installed in 2002. Inside the camera are electronic detectors, like those used in home video or digital cameras, which collect the distant light from stars. ACS has a wider field of view and better light sensitivity than the previous instrument and so immediately improved Hubble's viewing power by a factor of ten. Apart from cameras, there are also spectrographs on Hubble. These split light into its component parts allowing astronomers to work out a star's composition, temperature and age.

Edwin Hubble himself is probably now most famous as the man behind the space telescope's name, but he made a number of fundamental contributions to astronomy. Apart from calculating the distance to Andromeda and other galaxies, he discovered that the universe is expanding. Hubble realised that not only were galaxies moving away from us but that the further away they were, the faster they were travelling. That speed can be now be worked out by a formula known as Hubble's Law. To do this, you need to know the distance of the runaway object and a number called Hubble's constant.

The Hubble Space Telescope has taken incredible images of what are believed to be black holes at work. In 1998 it provided a unique view of a black hole in the centre of one galaxy gobbling up the contents of another nearby galaxy. Galactic cannibalism, as NASA put it. Hubble has given astronomers evidence of numerous black holes as well as insight into how they behave. The onboard Space Telescope Imaging Spectrograph (STIS) is crucial for this task. STIS combines a camera with a spectrograph, which splits any received light into the different wavelengths that constitute a rainbow. It can be used to determine what chemicals are in the swirling dust and gas and how fast they are spiralling into the black hole. It has been likened to a speed gun, as it is pointed at a suspect black hole to determine the rotational speed of anything travelling around it. The space police would certainly make money out of it because Hubble once found matter rotating arounda supermassive black hole in the centre of galaxy M84 at well over 1 million kilometres an hour. And, as every interstellar traveller should know, it's best not to speed in a built-up area.

By the way, if you've ever wondered why galaxies sound like motorways – M84, M31 and so on – they're named after the eighteenth-century French astronomer Charles Messier. He compiled an astronomical catalogue of more than 100 numbered objects and most of these turned out to be galaxies. If you read about galaxies that start with the letters NGC, this stands for New General Catalogue of Nebulae and Clusters of Stars. First published in 1888, this catalogue is based on entries by William Herschel. It now has more than 13,000 entries, of which 12,000 are galaxies. The existence of two catalogues explains why some galaxies, such as Andromeda, have two names: M31 and NGC 224.

LARGE BLACK HOLES (SIZE L)

Since we need a black hole to remove Slough, it's good to know that the search is not only under way but also extremely extensive. It therefore makes sense to consider the type of black hole we are likely to have to deal with in more detail. At the moment black holes appear only to come in two sizes: stellar mass or supermassive. They are measured in units called solar masses, which are, as the name suggests, based on the mass of our Sun. The Sun is therefore one solar mass. A star that is two solar masses has twice the mass of our Sun and so on. Incidentally, mass is not the same as size. Imagine that the Sun is a bath sponge for a moment. By squeezing the sponge into a tight ball, you have decreased its size, but its mass has stayed the same. This is because the contents of the sponge are still there; it's just that they now fit into a smaller space. So while a star that is 20 solar masses is still pretty huge, it is not exactly the same as saying the star is 20 times the size of our Sun.

Neither, incidentally, is weight the same as mass. Weight depends on gravity, but mass is constant. So although our own body mass, for example, remains the same no matter where we are, our weight varies according to the gravity of the object we are standing on. This is why astronauts are almost weightless on the moon, because although their mass doesn't change, the moon only has a fraction of the Earth's gravity.

Back to stellar mass black holes. These are formed when a large star of ten or more solar masses explodes and collapses in on itself. The resulting black holes are between 20 and 400 miles across and are anything between 5 and 100 solar masses. Astronomers have identified more than 25 stellar mass black hole candidates so far. One of the earliest was Cygnus X-1 in our own galaxy, the Milky Way.

Cygnus X-1, about 6,000 light years away, is considered one of the most likely homes of a black hole. It emits high energy X-rays and is in a binary system, which means there are two objects, or stars, orbiting around each other. The X-rays were first detected by the satellite Uhuru in 1972. The star gets its name from the fact that it was the first source of X-rays to be found in the constellation Cygnus, the swan.

The companion star (unimaginatively called HDE 226868) is huge, about 18 solar masses, and orbits its smaller, invisible, X-ray-emitting partner, Cygnus X-1, once every five and a half days. By definition, Cygnus X-1 could be a black hole because it has been calculated as having ten solar masses and also appears to be sucking in material from its giant partner. It is this accretion disc that is the source of the X-rays. These indicate that Cygnus X-1 must be small in size, which means that the mystery object is also compact and dense. Although accretion discs form around neutron stars, all the available evidence accumulated over the last three decades has led astronomers to believe that Cygnus X-1 is a black hole.

Not everyone was so convinced in the 1970s, leading to a famous bet between Stephen Hawking at the University of Cambridge and Kip Thorne from the California Institute of Technology. It concerned whether Cygnus X-1 was a black hole or not. Hawking said no; Thorne said yes. The stakes were four years' subscription to *Private Eye* for Hawking and one year's subscription to *Penthouse* for Thorne. Hawking conceded the bet in 1990.

In January 2001, a NASA scientist presented yet more information about Cygnus X-1 at an American Astronomical Society meeting. Almost a decade earlier the Hubble Space Telescope had observed ultraviolet light from clumps of hot gas fade, swirl and then disappear around the star. Exactly what you would expect around a black hole. The scientist, Joseph Dolan, spent years scrutinising all the data from Hubble, searching for the tell-tale event, describing it as 'looking for the proverbial needle in a haystack'. It was thought by many to be the first direct evidence of an event horizon (although not everyone agreed).

But Cygnus X-1 is not the best black hole candidate and hasn't been for almost 15 years. At the moment, first place goes to an object called GRS 1915+105. It has a mass 13–15 times that of our Sun and the argument for this mass is extremely straightforward: the 'invisible' X-ray source slings its partner star around extremely fast. It is, astronomers reckon, a black hole.

EVEN LARGER BLACK HOLES (SIZE XXL)

The other main black hole size is supermassive. These are a million to a billion times the mass of the Sun and are so large that one supermassive black hole could stretch across our solar system. Its accretion disc would, by virtue of its size, contain dust, gas and stars, and possibly whole solar systems. These are thought to be formed by huge gas clouds in the centre of a galaxy. Hubble has detected massive ones in the centres of several galaxies, including one called M87. It emits visible light and a jet of X-rays that have been photographed by Hubble and described as an extremely long blow torch. It is also surrounded by gas spinning around it at hundreds of kilometres per second and is thought to be around 2–3 billion times the mass of our Sun, which definitely puts it in the supermassive category.

England's current Astronomer Royal, Sir Martin Rees, and astronomer Joseph Silk suggested as long ago as 1970 that there were supermassive black holes at the centre of galaxies. Rees had already argued in 1971 that radio galaxies – galaxies that emit large amounts of radio waves – had active centres that spewed out powerful jets of energy. These ideas were later confirmed by radio telescopes. Astronomers then tried to explain how so much energy could be fitted into such a tiny core. All the answers seemed to point to a supermassive black hole.

Radio galaxies aren't the only potential homes for supermassive black holes. The brightest galaxies in the universe have extremely active centres emitting radiation and powerful jets. These are called quasars (quasi-stellar radio sources) and are also supermassive black hole suspects. In fact some astronomers believe that there are supermassive black holes in every galaxy, and recent research, published in the journal *Nature*, appears to support their claims.

The Milky Way

There are now an increasing number of people who have never seen the galaxy we live in. Light pollution from towns and cities is such a serious problem for both amateur and professional star gazers that it can permanently obscure an opaque band of stars stretching across the night sky. This is part of the Milky Way, an astronomical Catherine wheel whose four spiral arms contain dust, gas and over 200 billion stars. The galaxy we call home.

The Milky Way houses our solar system: planets Mercury, Venus, Earth, Mars, Jupiter, Saturn, Uranus, Neptune and Pluto; their moons; the asteroid belt between Mars and Jupiter and, of course, the Sun. The word 'Milky Way', like Mars and galaxy, has also been used as the name of a confectionery product (manufactured in Slough, as it happens) – we assume Uranus didn't make the short list. Edge on, the Milky Way (the galaxy, not the chocolate bar) does not look like the spiral that in fact it is. Instead, it resembles a 1950s-style flying saucer: a flat disc with a central bulge. The bulge is where the oldest stars can be found, as well the galaxy's suspected supermassive black hole. Our solar system lies towards the edge of the disc in one of the four spiral arms, called the Orion arm, so when you see the Milky Way at night you are actually looking at one of its other star-filled spiral arms.

All of the stars in the Milky Way, including our Sun, rotate around the galaxy's centre, with one revolution of the Sun taking around 225 million years – which gives an idea of the Milky Way's scale. It's huge. Yet it is just one galaxy among an estimated 50 billion in the universe. And there is more to our galaxy than meets the eye. It has a spherical halo which astronomers believe consists mostly of dark matter, so called because we don't know what it is yet. It was first proposed in the 1930s by Swiss astronomer Fritz Zwicky and is thought to be some sort of cosmic glue holding everything together. Scientists estimate that this missing dark matter makes up most of our universe so the search is on to find it.

In October 2002 a team of scientists from Germany, France, Israel and the United States announced the first indirect evidence of a supermassive black hole on our astronomical doorstep, located in the centre of our own galaxy, the Milky Way. The evidence was based on the behaviour of a star 26,000 light years away in the constellation Sagittarius. Radio emissions had previously been reported as coming from a source called Sagittarius A* (pronounced Sagittarius A star) in the middle of the Milky Way, making it a black hole candidate. By studying the orbit and acceleration of a nearby star, the team were able to conclude that Sagittarius A* was not a neutron star, a cluster of low-mass stars or even a stellar mass black hole but a compact object with several million solar masses. The conclusion: it must be a supermassive black hole.

Interestingly, because the X-rays from Sagittarius A* were faint, speculation surrounds the exact description of this black hole. One suggestion is that the black hole has exhausted all the surrounding matter in the accretion disc. Another is that it might be consuming the accretion disc's mass without radiating, which is why few X-rays are seen. Either way, the black hole in the centre of our galaxy is an unusual one. No one yet knows exactly how supermassive black holes are created.

EXTRA LARGE BLACK HOLES?

There is, as you may have noticed, a huge difference in size between stellar and supermassive black holes. If they were T-shirts this would be the equivalent of being available only in sizes L and XXL. So it sort of makes sense, from a T-shirt point of view, that some astronomers have proposed a new category of black holes in between the two sizes. It is the extra large XL T-shirt, or, in astronomical terms, an intermediate mass black hole.

Some astronomers have harboured suspicions that intermediate mass black holes exist ever since Dr Giuseppe Fabbiano of the Harvard-Smithsonian Center for Astrophysics identified some suitable culprits using the Einstein X-ray Observatory in 1989. More were found throughout the 1990s. They were called ultra-luminous X-ray sources (ULXs) because they appeared to emit so much light that the infalling matter would be blown away from a stellar mass black hole. The evidence suggesting that some of these ULXs are intermediate mass black holes, however, is not strong. If the X-rays from the ULX happen to be shining mainly in our direction, like the beam of a car headlight, the total emission is much less than if the X-rays spread out in all directions. In this case the ULX could have a normal mass a few times that of the Sun and still hold on to that infalling matter that powers the X-rays.

Years later the Hubble Space Telescope got involved. Astronomers detected two possible black hole candidates with masses of 4,000 and 20,000 times that of our Sun in two groups of bright stars known as globular clusters. The lighter of the two black holes was 32,000 light years away in the constellation Pegasus on the edge of the Milky Way, the other, heavier, one was in the Andromeda galaxy 2.2 million light years away.

Hubble tracked bunches of stars in each cluster and measured their velocities, as if measuring the collective speed of a swarm of bees. The astronomers found that the stars were being accelerated by objects that they determined were intermediate mass black holes. Their conclusions, however, have been highly disputed. But you can't keep a good idea down, and in March 2003 a team led by Dr Jon Miller (also from Harvard-Smithsonian Center for Astrophysics) presented further evidence for intermediate mass black holes. This time the European Space Agency's XMM-Newton satellite provided the data. It looked at two objects about 10 million light years from Earth in a spiral galaxy called NGC 1313, concentrating on the temperature of rotating gas in a possible black hole accretion disc. The results implied the existence of two black holes probably around 200–500 solar masses. However, other more conservative explanations with normal masses are again possible.

The available sizes of a black hole argument will probably run for a good few years yet but, at the moment, the strongest evidence is that there are only two sizes: stellar mass and supermassive (L and XXL). Up until recently it was believed that each galaxy contained only one black hole, but in November 2002 scientists at the Max Planck Institute of Extraterrestrial Physics found the

LIGO, LISA and friends

When two black holes collide – such as the ones heading for that merger in Galaxy NGC 6240 – scientists expect the resulting interstellar crash to produce gravity waves. These waves, predicted by Einstein in 1916, have no mass and will spread across the universe at the speed of light. They can be compared to those caused by a stone in a pond except that they cause ripples in the curvature in space. Scientists hope that both space- and ground-based observatories will be able to detect these ripples and in doing so will help discover how black holes are formed. The difficult part is detecting them, because gravitational waves hardly interact with matter and it has been said that they would alter the distance between objects situated on Earth and the moon by less than the width of an atom.

Nevertheless, the search is on. Ground-based detectors using giant aluminium cylinders have been around since the 1960s. Each cylinder weighs up to several tonnes and acts as a 'resonant-mass detector', the idea being that any gravitational wave will stretch the cylinder slightly as it passes through and will release some energy. Then, if the gravity wave is the right resonant frequency, rather like a tuning fork, the bar will ring or vibrate. One such detector, called Explorer, is currently in operation at CERN, the European centre for particle physics near Geneva. The main problem is isolating the cylinders from any noise or vibration so that it is possible to identify what causes any resonance. To avoid any contamination these types of detector are mounted on dampers, suspended in the air and cooled to near absolute zero (minus 273 degrees C) to remove thermal noise caused by heat. They are usually limited to detecting certain frequencies, and although there have been gravity wave claims by physicists using ground-based detectors they have never been confirmed.

One of the newest ground-based observatories on the block is LIGO, a Laser Interferometer Gravitational Wave Observatory. It links two large detectors in Louisiana and Washington State and became operational in 2000. LIGO is a joint project between the California Institute of Technology and the Massachusetts Institute of Technology. It was co-founded by Kip Thorne

(Caltech) and uses instruments called interferometers to detect any movement caused by gravity waves.

Interferometers are, as their name suggests, instruments that involve interference. Based on Michelson's designs of the late 1800s, the modern versions are extremely sensitive devices that examine the interference of laser beams. A laser beam is first split in two so that each beam travels at 90 degrees to the other before being recombined. The idea is that if one of those split beams is in the direction of an astronomical event, such as those colliding black holes, then the predicted gravitational waves will ripple across the beam and change its path slightly. When the beams join up again this minute difference will be reflected by an interference pattern (stripes of light and shade).

LIGO's first experiment ran successfully for 17 days in August 2002, but the results, published in April 2003, were negative. Nothing was detected. Scientists haven't treated this as a failure, however, as they were able to put some figures on how many gravitational wave events to expect in our galaxy – less than 164 per year for colliding black holes or neutron stars and no more than 1.4 explosions, such as supernovae, per day. Thorne wasn't too worried either. He had already predicted that gravity wave detection might not be possible until the second or third generation interferometers are built, with the required sensitivity. Money has already been set aside by the National Science Foundation for these future projects.

Space-based detectors are subject to less environmental noise in space and are therefore sensitive to lower frequencies. This is why another project, LISA, is under consideration. LISA, a Laser Interferometer Space Antenna, will be a joint mission between the American and European space agencies, NASA and ESA. It will consist of three satellites tethered together in space by a laser and will use the same basic detection principle as LIGO, but on a much larger scale. This laser would sense any changes in distance between the satellites caused by gravitational waves and would be directed towards events that are most likely to produce these waves.

first evidence for a pair of supermassive black holes in one galaxy. They were discovered 400 million light years from Earth by the Chandra X-ray Observatory in Galaxy NGC 6240. These black holes are about 3,000 light years apart, and they are circling each other at an incredible 22,000 mph. This speed will increase even further as they spiral closer and closer together. When they merge, in several hundred million years' time, gravitational waves are expected to ripple out across the universe.

Gravitational ripples were predicted by Einstein in his General Theory of Relativity but so far remain undetected – probably because they are so small that Einstein himself believed no one would ever be able to detect them. Advanced technology, however, may prove Einstein wrong, as it is hoped that observatories such as LIGO and LISA will confirm their existence.

Astronomers think the two supermassive black holes in one galaxy scenario resulted from two merging galaxies, each with its own black hole at the centre. They also believe that this may signal the future of our own galaxy as the Milky Way eventually collides with Andromeda, our nearest galactic neighbour. Don't stock up on groceries and resign just yet, as none of us will be around to see what's going on. This particular black hole merger isn't expected for another 4 billion years.

THE BLACK IN BLACK HOLE

No light can escape from a black hole, so you can understand how the black hole got its name. Yet, strictly speaking, a black hole is not black, at least not when you bring quantum theory into the equation. This is science on a subatomic scale where uncertainty and probability are rife and nothing is as you'd expect.

A black hole, on a quantum level, is predicted as emitting radiation. This radiation results from quantum uncertainty and the fact that, even in a vacuum, particles are winking in and out of existence. Pairs of particles – consisting of a particle and its antimatter equivalent, an antiparticle – are constantly being created and then destroyed. Hawking suggested that when these pairs of particles are created just outside the event horizon of a black hole, the strong gravity could suck in one particle before it had a chance to annihilate itself with its particle pair. The spare particle would then be propelled out of the black hole in the form of radiation. A black hole is therefore not black. It glows. This glowing radiation is now called Hawking radiation.

Hawking also suggested that another type of black hole might exist on a completely different scale entirely to those studied by most astronomers. He proposed mini black holes that are smaller than an atom. These, so far undetected, mini black holes formed during first few seconds of the Big Bang, the giant explosion that is thought to herald the origin of the universe around 15 billion years ago. The reasoning behind this is that the conditions would be exactly the same then as when a star collapses to form a black hole now and that the rapidly expanding universe may have caused some matter to compress into black holes. A mini black hole undergoing Hawking radiation would, in theory and over time, evaporate and shrink as its mass decreased. On disappearing from existence it would give off a massive amount of radiation. No one has even seen this, though, and there remains no evidence for mini black holes either direct or indirect.

BLACK HOLE FACTORIES

By this stage we know a lot more about black holes. Not the maths, thank goodness, but enough to impress anyone we meet at a party who is foolish enough to express an interest. So let's remind ourselves why we are looking at black holes. It is so that we can cleanly and effectively remove Slough, or anyone else's equivalent of an eyesore, and make the world a superficially better place.

So how do we make a black hole at home? We will definitely need more than an empty toilet roll and some sticky-backed plastic. At least that's what we thought until reading that an Irish inventor from Dublin claimed to have created a black hole in his shed.

In September 2002, Donal O'Gorman reported that he had artificially created a miniature black hole from radio waves by making them collide with

233

each other and disappear. This feat, it has to be admitted, was reported in a UK tabloid called the the *People* rather than a respected science journal and was dismissed in print by an unknown 'leading scientist' as preposterous. But, interestingly enough, scientists *are* researching acoustical black holes. In the 1980s Canadian scientist William Unruh suggested that sound waves in a fluid could behave like light in a gravitational field. If the fluid is moving faster than the speed of sound, then it would create an artificial black hole. He calls these sonic black holes 'dumb holes', as they are areas that can trap sound waves and … well … make them disappear.

As well as acoustical black holes, scientists are also investigating optical black holes, which trap specific colours of light. Researchers at Sweden's Royal Institute of Technology and the University of Scotland in St Andrews believe they can make these optical black holes in a laboratory. The only problem seems to be that you need a vortex of fluid reaching speeds close to the speed of light. Impossible? Maybe not, as there are substances such as rubidium gas or a Bose-Einstein condensate that can slow the speed of light down to only a few metres per second.

Needless to say, producing slow light is not easy. A Bose-Einstein condensate is an extremely cold blob of atoms produced by temperatures fractionally close to absolute zero – the lowest temperature possible (minus 273 degrees C). The blob results from the ability of atoms to behave like either particles or waves and the prediction that near absolute zero the waves will merge and condense together into a condensate – something that is neither solid, liquid or gas. This has been shown to happen in the laboratory, and scientists from MIT and Harvard have already demonstrated that by using two laser beams under these conditions, one of those light beams can be slowed down and made to stand still.

Dr Ulf Leonhardt at the University of St Andrews believes that if the laser beam is made to take the shape of a curve called a parabola and then stopped, the tip of this curve (think of it as the very front of a comet flying through space) will be the mathematical equivalent of a black hole singularity. The boundary will then be similar to an event horizon, and he will have created an optical black hole. In theory you would then be able to study black holes in great detail and observe Hawking radiation. Surely a Nobel Prize in the making for whoever gets there first.

Other scientists have suggested making the equivalent of a black hole using a type of helium called helium-3 that becomes a 'superfluid' at extremely low temperatures. In this condition helium-3 has no friction and doesn't behave like an ordinary liquid, not least because it will climb out of a glass. The link with black holes is a mathematical one because how sound travels through helium-3 is similar to how light travels in a gravitational field. If two of these superfluids are made to slide over one another, scientists hope they will be able to create a region where particles cannot escape the current and will thus have made a helium black hole.

Meanwhile, researchers from the Los Alamos National Laboratory are in the process of building what they describe as a 'black hole in a bottle' at the New Mexico Institute of Mining and Technology. It involves constructing a magnetic dynamo – a spinning magnet that will produce a magnetic field in the same way that the Earth's iron core spins and creates our planet's magnetic field, allowing us all to use compasses. The theory goes that there is an equivalent magnetic dynamo in a black hole accretion disc and, while no one fully understands how it works, by building one in the laboratory we can gain more knowledge about black holes themselves.

It must be said that all these methods are simply mimicking black holes. Acoustic, optic and magnetic dynamo black holes are simulations. Luckily for us, they do not have the accompanying gravitational field or we'd all be in big trouble. Our problem is that we want to assemble a black hole in its forceful entirety, just long enough to remove Slough but not long enough to destroy the Earth. Too large and its gravitational field will affect the tides, crack the Earth's crust and follow Jules Verne's fictional character, Arne Saknussemm, on a journey to the centre of the Earth. And then some. This is because the black hole consumes matter close to its event horizon in all directions, so it will gradually drop through the Earth until it reaches the other side. Gravity will then cause it to drop back down again, consuming matter on its way, but as the Earth will have rotated during this time, the black hole will be in a different position each time it reaches the surface. And so, rather like a pencil tracing the points of a star inside a circle, it will bore holes through the Earth until our planet resembles a Swiss cheese and is totally destroyed.

But the possibility of making a real black hole is not dead in the water. The Large Hadron Collider, or LHC, at CERN near Geneva, is a future black hole factory. Currently under construction, it will be the latest in particle accelerators. These machines are commonly referred to as atom smashers because they accelerate the particles that make up an atom to almost light speeds and then literally smash them together at enormous energies. From the air the buildings resemble giant concrete crop circles, and the LHC particle accelerator will be no exception. It will use a circular tunnel 27 kilometres long. When completed, in 2005, it will be the most powerful accelerator of its kind in the world.

Scientists hope the LHC will re-create the conditions immediately after the Big Bang and create miniature black holes. They also expect to witness Hawking radiation, as these mini black holes are likely to exist only for fractions of a second.

This then, is our best hope for removing Slough. We need to build a particle accelerator and create a mini black hole so that it exists long enough to snack on the town and then disappear out of existence before it does any serious damage. It is, we'll admit, somewhat risky.

Assuming that the 119,000 or so Slough residents have already been evacuated, we will need an event horizon that borders the town's officially designated 12.4 square miles (32.5 square kilometres). Then only everything within that event horizon will be stretched and squashed to pieces. This includes Slough's infamous trading estate (the largest in Europe), its Ice Arena (where the British Olympic ice skating champions Torvill and Dean used to practise) and, I'm afraid to say for any chocolate lovers, the UK home of the Mars Bar. But remember, things have to get worse before they get better, and soon all those badly

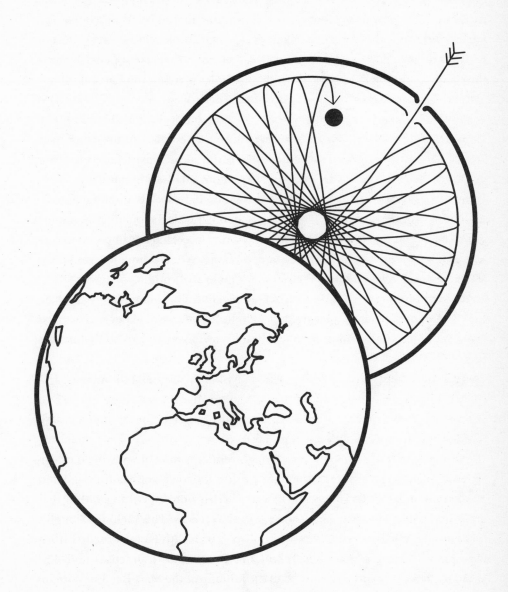

conceived buildings will be whirling around the black hole's singularity at millions of miles an hour before being crushed by its immense gravity. Then those brick monstrosities will never offend your architectural sensibilities again.

Any miscalculations and we are in big trouble. Extend the event horizon by a few miles and the black hole could accidentally wipe out London's Heathrow airport (just ten minutes drive away) and, since Heathrow serves more international passengers than any other airport in the world, immediately create a huge travel nightmare. Then there are the constitutional and historical implications if our mini black hole destroys nearby Windsor when the Queen happens to be taking tea in the royal residence of Windsor Castle with Princes Charles, William and Harry. Worse still, the corgis might be there too.

Before there's a sudden drop in property prices in Switzerland, we must stress that the Large Hadron Collider near Geneva will be perfectly safe. Only our version might be dangerous. Can it ever be done? Maybe. If the LHC delivers a mini black hole and perhaps dark matter is involved to help keep things under control. But by the time a particle accelerator is built on Slough's trading estate, for instance, the whole town will be aware of our plan. The town council will then, no doubt, embark on an intensive plan of rebuilding and landscaping and begin filling the streets with hanging baskets. So the very suggestion of building a miniature black hole to wipe out an eyesore could well be all we need to produce improvements.

But perhaps the best use of a black hole we have come across is from Professor Fred Adams at the University of Michigan. He kindly came up with a way of using black holes to remove eyesores that is far less likely to destroy the Earth. He believes the world could benefit from a black hole disposal unit in orbit. It would need to be at one of the so-called Lagrange points in the Earth-Sun system, where the black hole can orbit and everything in that system will remain stable. It must therefore have a mass much smaller than both objects, approximately that of Mount Everest. A black hole with this mountainous mass would be smaller than an atom, but it would have a sphere of influence about 160 metres across. The idea would be to fly a space shuttle or other craft to within 100 metres of the black hole and dump any rubbish into it. As this guzzling black hole garbage dump would generate energy, solar panels could be placed around it to keep the Earth from being zapped by high energy photons and also act as a generating power source. Two products for the price of one. The black hole would then chomp away at life's detritus, and the Earth would never again be blighted by all those landfill sites and rubbish dumps. Now that's what we call a result.

CHAPTER EIGHT

How to Live For Ever

All things considered, there's only one certainty in life. That you'll die. But what if science could change that?

In one lifetime you can do only so much: a bit of education, a dozen or so partners, a couple of kids, a few jobs and some nice holidays. Along the way our hearts will beat around three billion times, we'll watch at least three years of television and spend 100 days having sex (sometimes at the same time). We might be amazingly successful, wealthy and glamorous or just muddle through. But however you have lived your life and whatever you have achieved, the outcome's the same. You die.

But what if life wasn't restricted by biology? What if you could live for 100 years, 200, 500, 1 million, for ever? Think of all the things you could do, the money you could make, the places you could visit, the sex you could have.

At the moment the average human life expectancy is about 65 for men and 70 for women. Japanese women live the longest, to an average age of almost 85. The British don't fare nearly as well. On average women die at 79, men at 75. In the short term there is some good news. A recent study by scientists at Cambridge University suggests that people, in the Western world at least, are living increasingly long lives, and in a few decades reaching 100 will be completely unexceptional.

But it's not for ever is it?

Living to 100 is all very well, but when you get to that age, frankly you're going to look it. There's not much point living for ever if you can't enjoy the benefits of a new body every so often. More importantly, you need to keep your mind in shape; there's no benefit in extending your lifespan if you can't remember who you are. Of course, there may or may not be an afterlife, but even if you believe there is, don't you sometimes think it would be nice to remain in *this* life just a little longer?

WHO WANTS TO LIVE FOR EVER?

We'll be straight with you. Living for ever is not going to be easy. To begin with we'll set our sights reasonably within the bounds of reality. There are a number of scenarios, some more probable than others, that might help you realise the dream of never-ending life. It's just that scientists haven't worked out a lot of the detail.

The problem can be broken down into two areas – the body and the mind. The great thing about the body is that scientists more or less know how it works. From how the circulation of the blood delivers oxygen to the organs, to the chemical processes within each individual cell. What this means is that science will ultimately produce a means to keep the body alive for longer, whether that means treatment with gene therapy, an upgrade with cybernetics or the application of nanotechnology.

The cells in our bodies are constantly being replaced. Look at your hand for a moment. It's a completely different hand from the one you had a few years ago. Even though it looks the same and it's still your hand, almost all the cells have been replaced. Clearly this doesn't diminish you as an individual – you're still the same person with the same personality and sense of self. What remains unaltered throughout is your genetic blueprint. When one cell dies, another genetically identical cell is created to replace it.

But even losing some of that unique DNA wouldn't make much difference. There are thousands of people walking around who've had bits of their bodies replaced with parts from other people. Heart, lung, kidney and bone marrow transplants all introduce genetically alien cells to the body. Or consider the cybernetic technology of Chapter 6. In most cases the recipient isn't any less of a person.

So what happens to our body isn't so important; the secret of living for ever is to find some way of preserving our mind, our conscious self. That might mean downloading it into another body, transferring it into a computer or building an artificial brain to house it in. But before we even attempt to grapple with one of the most fundamental problems of science – what this thing we call consciousness is anyway – there is a more immediate concern. Time is running out.

PUTTING THE PROBLEM ON ICE

Every minute of every day brings death a little closer. Frankly, at the moment, scientists have got quite a lot on their collective plates (or petri dishes), and getting round to preserving consciousness is not top of anyone's list. So our immediate priority is to buy more time until they figure it out. At the moment the best available technology is not so much life preservation as just preservation.

Cryonics is the process of preserving humans. It's experimental, unproven and the humans must be dead first. Proponents view it as a final safety net. Critics dismiss it as another pseudo-science. However, unlike many of the areas of research in the hinterlands of science, cryonics does have a basis in scientific reality. Before we go any further, please note that blind optimism is also a crucial requirement.

The idea of cryonic suspension is to preserve an individual at a sufficiently low temperature to prevent the deterioration of cells. The hope is that future generations will be able to revive the person, allowing them to live again. Of course, future generations might have better things to do than bring back the dead.

Cryonics is based on the principal of cryogenics, the science of the very cold. Low temperatures are already used to store embryos for human fertilisation programmes and the genetic material of endangered species in DNA banks. The science behind all that research is certainly sound and we know

Legally dead

Apart from the very practical problem that it might not work, there are a number of sticky legal issues standing in the way of a cryogenic future. The most difficult to pin down is what exactly is meant by death. The definition can vary from country to country – and in the US from state to state.

Broadly, someone is said to be legally dead when their body stops working and there's nothing else medical science can do about it. Today this generally means 'brain death', although even this definition is open to varying interpretation, depending how much is up for grabs in the will. Deciding what death means also depends on the prevailing medical knowledge. One hundred years ago people who stopped breathing as a result of drowning were often declared dead. Today, if they are reached quickly enough by someone trained in first aid they're likely to survive.

At the moment the cryonics industry is based in the United States, which makes it subject to that country's laws and regulations. Because cryonics isn't defined in law in the US as a 'lifesaving procedure', any 'patient' has to be dead before being frozen. If that ever changed, people could be cryogenically frozen while still alive, better ensuring the body's integrity. The organisations involved in cryonics would very much like to see this procedure adopted, but in the meantime they try to reach their clients as soon as possible after death. By their definition, the people they freeze are still 'alive'; it's just that at the moment medical science hasn't developed the technology to treat them.

As to the legal definition of 'brain death', there isn't one, not in the UK anyway. Instead there's an accepted practice that is followed, but it's not strictly speaking written in law. A person is taken to be brain dead when the brain stem completely ceases to function. This is the small area of the brain at the top of the spinal cord responsible for, among other things, the automatic control of breathing. If the brain stem is dead then there's no way the patient can breathe without a ventilator and neither are they able to regain consciousness. The hospital procedure to make sure someone is truly dead involves a series of tests carried out by two different doctors. For obvious reasons, neither one of the doctors can be a transplant surgeon.

But like most things in life, death does not always go to plan. There are a couple of other types of death that are not so clear-cut. First there's suspended animation, which can be brought on by drowning or extreme cold, where the patient to all intents and purposes is dead but can be brought back to life. Then there's PVS – persistent vegetative state – where the brain stem is working but the individual shows no signs of being aware of anything. It certainly appears as if their consciousness is no longer functioning and they have lost any sense of 'self'. This poses a particularly difficult dilemma for both family and medical staff: should they 'allow someone to die' or wait. People with PVS have been known to recover.

What cryonics practitioners want to avoid is what they call 'molecular death', where circulation ceases completely and all the cells in the body shut down. They aim to get to their patients before that happens or at least before the majority of cells in the body have packed in. Which gives them about an hour after death. You might, of course, die in your sleep at ten at night and not be discovered until the morning. In which case it's probably too late.

You might also want to consider the tax implications of dying. In the UK, for example, until cryonics is accepted as a medical procedure, you would still be liable for inheritance tax (providing your estate is worth a sufficient amount). Spending a year dead for tax reasons, a technique employed by Hotblack Desiato in *The Hitchhiker's Guide to the Galaxy*, isn't yet an accepted tax dodge. Sorry.

that embryos and DNA can survive the process. The question is, can the rest of a human being?

Cryonics feature in a number of science-fiction books and films, notably, Woody Allen's *Sleeper*, in which a neurotic New York jazz musician (that'll be Woody Allen then) goes in for some routine surgery and wakes up 200 years later to a world of giant vegetables, domestic robots and orgasmatrons (which do exactly what they say on the tin). More recently, Matt Groening's cartoon series *Futurama* is based on a pizza delivery boy (Fry) who is accidentally cryogenically frozen in December 1999 only to wake up 1,000 years in the future.

In both *Sleeper* and *Futurama*, the heroes are out of synch with the rest of society – with, as they say, hilarious consequences.

So how can it work for you? First you need to sign up with one of the companies offering the service. Remember, when you're dead, it's already too late. You are looking at a long-term investment, so you want to ensure that the company is on a firm financial footing. If anyone fails to pay the electricity bill, for instance, you're a messy puddle on the floor. One of the largest organisations involved in cryonics is the Alcor Life Extension Foundation, which has more than 50 individuals currently preserved at its facility in Arizona. Although the methods each company uses vary slightly, the basic principal is the same.

REBORN IN THE USA

Once death is pronounced, the aim is to get the corpse (sorry, patient) preserved as soon as possible. The patient is lowered into an ice bath and, with the use of a resuscitation machine, their circulation and breathing are artificially restored (although they are still, of course, technically dead). The blood of the patient is then gradually replaced with antifreeze, not dissimilar to the stuff you put in the car radiator in winter. Once this process is complete, the patient is immersed in alcohol to be cooled down even further. There are worse ways to spend a few decades.

It then takes two weeks to chill the patient to the temperature (minus 196 degrees C) of the liquid nitrogen in which they are eventually immersed for storage. Or rather, 'long-term care', which basically means someone keeps paying the electricity bills. Patients can just opt to have their heads preserved, the logic being that their bodies weren't that much good anyway and science will come up with something better. Heads are stored four to a jar, so at least they have some company.

This is all relatively straightforward and arguably a good way of storing the bodies (or heads) of dead humans relatively intact. The big question is, how intact? If all goes to plan and the preservers get to the dead individual soon enough, then they will have acted before all the cells die. There will almost certainly be tissue damage from ice formation, because water expands as it freezes, rupturing fragile membranes. The cryonics companies claim that the slow cooling process and the antifreeze should minimise this. The DNA will be preserved, but as we have already discussed in Chapter 1, that cannot be used to re-create exactly the same person, only a physical copy. To be successful, cryonics has to preserve the mind of the dead. Of course, the individual concerned must have died of something, so there is probably something wrong with them.

Cryonics relies on future medical advances. If a frozen embryo can be restored to 'life' runs the argument, surely in future so can a whole human being. Much is made of advances in nanotechnology. Pioneers of cryonics predict that machines 1,000 times smaller than a cell will be able to repair any damage from the inside, each cell being re-engineered to function again. Crucial to this will be the ability to restore the function of the brain, the personality and memories. Whether this can ever be achieved depends very much on the nature of the mind.

All cryonics companies say they are setting sufficient funds aside to revive their patients if the technology becomes available. But would wider society want them to? At the moment the world population is around 6 billion and it goes up by about 170 every minute. By 2011 it will be 7 billion, by 2035 around 10 billion. Almost double. The last thing the world will need is a load of 'reborns' from the early twenty-first century. Will the prevailing laws of the time even allow the dead to be unfrozen? How will they fit into society and what use will they be?

The first to be successfully unfrozen might be a bit of a curiosity and fascinating for historians and daytime TV audiences, but what about the rest? The chances are they will be physically and mentally old and a burden upon their great, great, great-grandchildren. Also, any people reborn in, say, the year 2100, will to their minds have travelled forwards in time. They will no more fit into society than a Victorian lord would fit into ours. They would most likely be terribly lonely. Assuming they died at a late age (in their eighties, say), they would just be getting to grips with everything when they die again – and presumably have to be frozen again for another leap through time. Even if cryonics works, and the jury's still out on that (in fact the jury's just getting

Frozen Dead Guy Festival

Some odd things happen in small-town America. But Nederland, near Denver in Colorado, is stranger than most. Every year it plays host to the Frozen Dead Guy Festival, which features coffin races and a Grim Reaper parade in honour of the town's most famous resident: 'Grandpa' Bredo Morstøl. Who is dead.

Bredo was a Norwegian who died in 1989. How he ended up frozen in dry ice in a shed in Colorado is a story that is, not surprisingly, somewhat bizarre. His grandson, Trygve, had lived in the United States since 1980. Something of an eccentric (not that there's anything wrong with that) he, in no particular order, founded a club devoted to bathing in icy water, was arrested after joking about hijacking an aircraft and built a house capable of resisting nuclear attack. He was deported in 1994 when his visa expired.

At this point it emerged that the body of his dead grandfather was being stored in ice in a ramshackle shed behind his house. Trygve had had it shipped from Norway shortly after death. Nederland held an emergency town meeting but no one could find any law forbidding the preservation of dead bodies. A new ordinance was passed banning future retention of corpses but grandpa Bredo was left alone.

A new shed was donated by a company called Tuff Shed and a small industry has grown up around the frozen dead guy. Along with the festival, you can book tours of the shed or view the documentary film *Grandpa's in the Tuff Shed*. The man himself is wrapped in dry ice, replenished every month by a group of planetary geologists (well, who else?). Grandpa Bredo's views on cryonics are somewhat unclear – he'll presumably make them perfectly plain when he's revived.

As for Trygve, he hasn't returned to the United States, although the ice-water swimming club he founded now boasts 350 members. He is still involved in cryonics and has an interesting website detailing his views. He's a big supporter of the Frozen Dead Guy Festival.

comfortable in its hotel, ordering delivery pizzas and pay per view porn), it's not so much a way of living for ever as a method of seeing the future.

However, the ability of cryonics to effectively buy time could find other uses. If scientists can perfect the freezing and subsequent thawing of humans, it would be the perfect way to travel the vast distances involved in interstellar travel. This is a device employed in many a science-fiction scenario. Astronauts are frozen for the journey to be reawakened years later on the other side of the galaxy. Although the travellers are not technically dead, a freezing process has slowed their metabolism right down, meaning that they age at a far lower rate.

With the exception of faster than light travel, cryonics is probably the most practical way for humans to travel across space. Even a trip to Mars currently takes at least six months. How much better if that time was spent asleep or clinically dead rather than bickering with your companions. There would also be a considerable weight saving, with the need to carry less food and water on the trip. The author and space visionary Arthur C. Clarke uses the idea of suspended animation for astronauts in a number of his novels, notably *2001*, in which the artificially intelligent HAL-9000 computer kills all the frozen astronauts. So if you're planning to sign up with a cryonics company, just check they're not using the HAL series of computers.

As for living for ever, cryonics might provide a stop-gap, but it doesn't really solve the underlying problem, merely postpones it. Crucially it relies on the body being in a decent state to start with. A body riddled with cancer and containing a brain decimated by Alzheimer's is going to do no one any good. While cryonics is a reasonable idea in theory, it might prove impossible in practice. Take a look at the small print. No one involved in the business is making any guarantees.

DOWNLOADING THE MIND

Optimists see cryonics as a way of ensuring that they can live again. Others see it as a leap of blind faith in future medicine and society. Even if medical science is able to revive a body after 100 years, restore each cell to full working order and get the circulation up and running, that doesn't necessarily mean it will be able to restore the person. What effect does dying have on the consciousness? Is the mind something that is preserved in the hard wiring of the brain, in the connections between cells? Or is it something much more subtle and nebulous that needs more than the physical structure to survive? What about memories? Is there any point living for ever if you can't remember who you are or what happened to you in the past?

Several futurologists and even scientists have suggested that it might be possible to download the brain onto a computer. This would be a way of preserving memories, personality and the consciousness of the individual without having to trust to cryonics. Professor Hans Moravec, a senior American robotics researcher has even thought about how it might happen.

He imagines a future in which a human might inhabit an artificial reality to the exclusion of the real world. To exist in a virtual world a human being would not need a body any more, just a brain. If they wanted to interact with the real world they would do so through a robot. The rest of the time they would exist in a computer-generated reality. It's an image he describes as a 'brain in a vat', fed with all the stimuli and nutrients it needs but disconnected from the restrictions of physical reality. Over a period of time, like all brains, it will start going wrong – mental agility will be lost, memories forgotten. So the technicians start to replace the malfunctioning components with electronics. Until eventually the whole brain is replaced.

Professor Moravec assumes that in doing this, consciousness, personality and awareness will remain unaffected. If the process is successful, the result is that a computer has replaced the brain and yet the person lives on, in silicon. Taken one step further, what if the downloading could take place without replacing the brain? Instead, a process could be devised that involved copying every individual neuron and connection. Critics, notably Rodney Brooks, Moravec's former colleague, counter that the idea of copying the brain's cell structure neglects the role played by the chemicals that regulate brain activity surrounding each and every cell. Nevertheless, it's an intriguing thought and it's not inconceivable that in future computers will be powerful enough to cope with the contents of our heads.

WHO AM I?

So far in this chapter, we've avoided the fundamental underlying question – what is this thing called consciousness anyway? What makes each of us a person? The brain is a wet, greyish, 1.4-kilogramme lump of 100 billion nerve cells, or neurons. Somewhere within that tangle are our thoughts, memories and individuality. Somehow we have evolved a consciousness – an awareness of what we are doing and how and why we're doing it. We can think, therefore we are.

'What is consciousness?' is described by American Philosopher David Chalmers as 'the hard problem'. This is something of an understatement. It's one of the great unknowns, so much so that scientists don't even know if it's the *right* question. By that we mean they don't really know where to start. Nor even which branch of science is best equipped to come up with a solution. There's something going on in our mind that enables us to have subjective thoughts and an awareness of our surroundings. But what *is it*?

It's perhaps easier to think of what consciousness isn't. Why aren't we all walking round like robots or zombies, just going through the motions of existence? How have we evolved the ability to question why we're here and what's going on? And does this make humans unique or do other animals have some degree of conscious thought. If not, are Fido and Kitty mindless biological machines?

Chalmers divides 'the problem' into two – what he calls the 'easy' part and the 'hard' part. The easy problems of consciousness are things such as whether it's too hot or cold and how much something hurts. These can be explained through known physical 'pathways' in the brain. The hard problem is one of experience – how do we feel and think? Not everyone agrees with Chalmers, but his formulation is as good a place as any to start.

There are broadly two schools of thought: those who see consciousness as some mystic quantity beyond our comprehension and those who believe it's something occurring as a result of physical processes in the brain. There are

Inside your head

The brain is not perhaps the prettiest thing to look at. It resembles a sickly greyish walnut about the size of a kid's football stuck on the end of a stick. Although many of its workings are still deeply mysterious, scientists have got a pretty good idea of which particular areas do what.

If you look at it head-on, as it were, each side looks identical: the two deeply rutted hemispheres are called the cerebral cortex. We'll come back to them in a moment. Underneath the cerebral cortex, at the back, are the cerebellum and the brain stem. These are right at the top of the spinal cord. The cerebellum looks a bit like a miniature brain, which in a way is what it is. This is the area concerned with the mechanical and unconscious. It has the same components as a reptile's brain and does much the same things. Attached to the top of the brain stem is a part of the brain called the limbic system. This has only evolved in mammals and consists of a whole load of interconnected modules. The limbic system is where emotions are generated. It's also where you'll find the hippocampus, the place where long-term memories are formed.

The really interesting stuff happens in the cerebral cortex. Each of the two hemispheres is divided into four lobes. Even though your eyes are at the front (unless you're a schoolteacher) it's the back lobe, the occipital lobe, that deals with vision. Then there's the temporal lobe. This processes hearing, perception and emotion – what you feel. The parietal lobes at the sides of the brain create a three-dimensional view of the world for your mind and place you in it.

Most of what we consider to be consciousness is in the frontal lobes, behind our forehead. This is where our brain generates our sense of 'self' and where we consciously 'think' about things. There's a big difference in the size of this area of the brain between humans and our evolutionary ancestors. This seems to be the bit that makes us human. The frontal lobes are connected to most other parts of the brain, but aren't essential for keeping us alive, just for us having any sort of life.

Messing with the frontal lobes of the brain used to be a particularly lucrative source of income for jobbing brain surgeons. At a conference in 1935, Portuguese surgeon Antonio Moniz saw a demonstration of the effects of removing the frontal lobes from a chimpanzee. As a result he went on to develop a new 'treatment' for mental illness, the lobotomy – destroying nerve tissue to cure the mind.

Certainly, to the surgeons who went on to adopt Moniz's techniques, their patients appeared to show signs of improvement. They were no longer so agitated or paranoid. Instead, they were apathetic and tended to suffer from nausea and disorientation. But that's medical progress. By 1949 surgeons were performing 5,000 lobotomies each year in the US alone. In the same year Moniz won the Nobel Prize for his contribution to physiology and medicine. Apart from the awful side-effects, lobotomies were not entirely successful – Antonio Moniz ended up being shot by one of his lobotomised patients and later died as a result. Which has a certain justice to it.

Although this operation is thankfully a thing of the past, surgery is increasingly being used to help cure brain disorders. Only this time it's a lot more precise. In most cases, though, drugs are a far more effective alternative.

quite a few theories about the latter, and (since 'beyond our comprehension' is by definition incomprehensible) we're going to concentrate on those. One of them might even be right.

Underlying all these theories is another difficulty, known as 'the binding problem'. Our subjective experience is made up of a whole lot of different perceptions. Stop reading and look around for a moment – your mind is receiving information from your eyes, ears, nose; you're maybe thinking about your next meal, contemplating the meaning of life or wondering how you can keep reading this sentence while doing all that. All those thoughts are being processed in different parts of your brain. Somehow your consciousness is able to bind them all together.

The scientists who approach the problem of consciousness at a totally physical level are for the most part neurophysiologists. They study individual neurons (nerve cells) and groups of neurons to work out how they function and interact. By understanding these processes they hope to explain the behaviour of the entire brain. The interactions between neurons are certainly extremely complicated. Francis Crick, the co-discoverer of DNA, believes that everything from our free will to our individual identity is nothing more than the behaviour of a brain full of nerve cells and molecules.

An alternative view is that all the complex interactions between neurons lead to the emergence of a higher level behaviour (see Chapter 2). In the same way that the behaviour of an ant on its own is very simple, the emergent behaviour of an ant colony is extremely complicated. Although each unit in the colony has a very limited set of behaviours, together ants can achieve remarkable things. Likewise, neurons are relatively simple units on their own but millions together make a brain. Perhaps complex behaviours such as consciousness emerge as a result.

Although neurons are the basic building blocks of the brain, other scientists take the view that the interactions between them, even sufficiently complex ones, don't fully explain the nature of the mind. They look instead to the weird world of quantum physics, where everything is viewed at a sub-atomic level and common sense goes out of the window. They contest that consciousness can only be explained by quantum effects taking place within each individual neuron.

Then there are the psychologists, who take a broader view. Rather than examining the minute biology of the brain, they attempt to explain consciousness through the way we behave. As we'll see, their experiments sometimes reach disturbing conclusions. Finally, there are philosophers, who, despite centuries of thinking about the problem of what consciousness is, still haven't come up with a solution. But as no one else has either, they're still on to a winner.

Consciousness could, of course, be a combination of all these theories. Obviously, if we knew what it was, we wouldn't be writing science books but living off the spoils of our Nobel Prize. It's worth pointing out that there is considerable overlap between research in consciousness and research into artificial intelligence. Many scientists have a foot in both camps. After all, if AI researchers manage to build a conscious machine, in so doing they will hopefully come up with an explanation of consciousness. For our purposes, if we want to live for ever, understanding consciousness is important because we need to know how to preserve it.

Now pay attention

Imagine you are driving down the motorway on your way to a job interview. The radio's burbling away and you are on the phone arguing with your partner over the state of the bathroom. You are obviously conscious as opposed to unconscious, but how aware are you of what's going on? You might think you are concentrating on your driving, but when you *really* pay attention to your driving, you end up trying to figure out how you got to where you are. You can't remember being aware of the road at all. Different things bob up to the conscious part of your mind: the bathroom, the irritating DJ on the radio, worries about the interview, back to the driving. Most of the time you're doing at least one thing on autopilot while consciously doing something else.

What we perceive to be our consciousness is actually a whole load of different perceptions clamouring for our attention. Researchers refer to these elements of our conscious experience as 'qualia'. At any particular instance one particular qualia becomes more prominent than the others, the next instance another might, and so on.

As for our perception of reality, that's all over the place. Dr Daniel Simons and his then colleagues at Harvard University carried out a series of experiments that demonstrate a phenomenon called change blindness. In one experiment, a researcher approaches someone to ask for directions. While the researcher and the unsuspecting subject are talking, another couple of experimenters barge in between them. During the interruption, someone who looks completely different replaces the original researcher. Around half the subjects completely failed to realise they were now talking to a different person.

Another experiment carried out at Harvard demonstrates a similar phenomenon called 'inattentional blindness'. It involves asking subjects to pay attention to a basketball game. During the game a man dressed in a gorilla suit walks in front of the players. Most subjects watching the game completely missed the gorilla.

What experiments like these demonstrate is that we do not record reality in the same way that a video camera does. What we 'see' in a particular situation competes in our brain with memories of what we've already seen and expectations of what we are going to see next. The brain constructs an illusion of what is going on. It's usually right but not always. There's not a direct hard link between the area of our brain that deals with vision and the part that controls our consciousness; other stuff gets in there too.

BLIND AMBITION

More remarkable still is a syndrome called blindsight. It was discovered by Professor Larry Weiskrantz at Oxford University and has been cited by consciousness researchers ever since. To appreciate it, you need to understand how sight works.

When we look at something, the image formed on the retina of our eye is sent to two different parts of the brain. It goes to an area in the brain stem (the most primitive part of brain) and to the more recently evolved visual cortex (part of the occipital lobe of the cerebral cortex). When the visual cortex is damaged, you become blind. Because the brain does things back to front, if the right visual cortex is damaged, you are unable to see anything on the left side of your nose and if the left visual cortex is put out of action, you lose the sight on your right. Remember, the eyes are still functioning perfectly normally; the brain has just lost the ability to process the signals.

Professor Weiskrantz studied a patient code-named GY to protect his identity. GY was asked to point to an object in the 'blind' region of his vision. This, obviously, seemed to him a somewhat bizarre request. He eventually did as he was told and reached out and pointed to it exactly, even though he could not see it. GY claimed he was guessing where the object was, but in 99 per cent of trials he got it right, even though he could not consciously perceive what he was being asked to point to. So what is going on? Is this proof of extra-sensory perception? Remember, there are two visual pathways to the brain, one to the primitive part of the brain and the other to the conscious brain. GY was unconsciously 'seeing' the object, but the perception was not making it to his conscious brain.

This suggests that one part of the brain is conscious and another unconscious and that some things break through to our consciousness and others remain buried somewhere in the depths of our brain. Dr Vilayanur Ramachandran, one of the world's leading figures in consciousness research, uses blindsight to explain the driving example we mentioned. He contends that we often drive using blindsight. This is why we can be seemingly unaware of what we are doing. We can carry on conversations or think about the state of our bladder because visual signals aren't making it to our consciousness.

Others disagree with that assessment, but then that is the great thing about this area of research: no one has the answers. So where does this get us other than very confused? What it suggests is that if we want to look for consciousness we will probably find it in the parts of the brain that have evolved more recently. Sadly, it does not give us a definition of consciousness.

SELF-CONTROL

Here's another weird thing about the brain. We are under the impression that we are in control of our actions. As we write this book, for instance, we are thinking about what we're typing before we type it. It's common sense. However, the author Enid Blyton once told a psychologist that her characters emerged in her mind and she wrote them down. She did not really think about them or work them out in her consciousness as ideas. This certainly explains Noddy and Big Ears. Dr Steven Pinker, a Professor at MIT who studies the evolution of language, has a simple explanation for this. He believes that we don't usually talk or read consciously at all, except when we're grappling with something particularly difficult. He likens human speech to a spider spinning a web. Just as the spider doesn't reason through the intricacies of web geometry, we don't consciously think about what we're saying.

This is backed up by research done by Benjamin Libet at the University of California. He studied patients who were undergoing brain surgery. During

A magnetic personality

A new theory on consciousness is being championed by a professor of molecular genetics at the University of Surrey. Johnjoe McFadden calls it the 'consciousness electromagnetic information field theory', or cemi field theory. It's based on the electrical properties of the whole brain.

The brain is made up of a network of neurons, each transmitting electrical signals along nerve fibres. Although at a microscopic level their workings are incredibly complex, at another level the brain can be considered little more than a tangle of badly labelled wiring. When electricity passes along any wire, it generates a field of energy, known as an electromagnetic field or EM field. These EM fields are all around us, from TV and radio signals to the energy fields generated by our computers or household wiring.

An EM field contains the same information as the circuit that produces it. So in the brain, the field surrounding a particular neuron reflects the nature of the electrical current coursing through it. Like any energy wave, fields can combine with others or cancel each other out when they interfere. We know that the brain as a whole generates its own EM field; what McFadden proposes is that these waves of energy encompassing our brain are the seat of our consciousness.

What's his evidence? Well, for a start no one can prove him wrong. But the most compelling argument in his favour is something called 'neuronal synchrony'. In experiments in which a subject recognises a face, a whole load of neurons fire together. As a result, each neuron generates an EM field. Because these fields are synchronised, the energy waves combine together, reinforcing each other to generate an overall disturbance across the brain. This, McFadden believes, corresponds to our conscious awareness.

If this is the case, it would have to also work the other way round. Our consciousness has to be able to set off neurons in order for our conscious instructions to be carried out. McFadden concedes that an EM field cannot trigger a neuron to fire from a standing start, but it can set a neuron off when it's on the *brink* of firing. This, he says, means EM fields come into play 'when the brain is poised to make delicate decisions'.

The cemi field theory certainly solves the binding problem, explaining how our consciousness can co-ordinate activities across disparate parts of the brain. It is also perhaps an explanation for free will: the cemi field is what makes us able to make decisions. But if proved correct, the theory will create some serious problems for AI researchers, suggesting that the only way to achieve a thinking computer is to copy exactly the functioning of the brain. Which rather scuppers our plan to download our consciousness into a conventional computer. However, says McFadden, all is not lost: 'It may be possible to design a computer that works in the same way as the brain.'

the operation the brain is exposed but the patient remains awake. Scientists can prod around without the patient feeling a thing because the brain does not have any pain sensors. He stimulated areas of the brain that correspond to different parts of the body. For example, touching one spot in the brain would make the subject feel a touch on the hand. Every time he stimulated the brain, it took half a second for the subject to register the touch in their consciousness. When he stimulated the hands directly, the same delay was experienced.

This means that our consciousness is registering what is going on half a second *after* it happens, which suggests the idea that we are living in the present is an illusion. Likewise, the impulse in our brain that tells us to do something happens half a second *before* we do it. We can choose to stop the action, but the action is already started. This has led some people to claim that our consciousness does not exert any control over what we are doing – that we cannot make our own conscious decisions; instead, consciousness itself is an illusion created within the brain. This would explain why scientists can't understand it, because it doesn't really exist. Or at least not in the way we think it does.

Psychologist Susan Blackmore suggests that the brain is concocting a series of stories about the world. So every time you probe your consciousness it puts together a retrospective tale of what you are doing. It also gives you the feeling of 'self' – that *you* are the one experiencing it, not someone else. Until you asked your mind what was going on, she suggests, there was no stream of consciousness, because there is no such thing. There just seems to be. It's only when we bother to ask something that a story is concocted.

If this is true, then scientists are asking the wrong question. Instead of trying to figure out how electrical signals get turned into conscious experiences, they should be asking how the illusion of consciousness is constructed. As for the idea of downloading your consciousness onto a computer, if it doesn't exist as a separate entity there's no need to worry about losing it. Once revived, a preserved brain with all its memories intact would start generating its own consciousness again.

CONFUSED?

This is all quite possibly making your brain hurt. However, the great thing about consciousness is it makes even the greatest brains hurt as well. No one knows the answers; at the moment scientists are just working on different approaches to the question. Oxford pharmacologist Professor Susan Greenfield is looking at the problem from a more physical perspective.

She believes that there can be different levels of consciousness. This would seem to make sense. A raw emotion like anger feels very different from inner reflections on the meaning of consciousness. We can raise or lower the amount of conscious thought we devote to a particular thing. She suggests that consciousness is not confined to any particular centre of the brain but occurs when ordinary brain cells form networks. These vary in size depending on the depth of thought.

She points to studies that show that as many as 10 million brain cells network together in response to a single flash of light. Alternatively, when we are dreaming there are only small assemblies of cells, so the storyline becomes disjointed. Drugs like ecstasy disrupt the chemical connections between neurons, perhaps affecting the ease with which these hubs of cells form.

For this theory to work, there would have to be something to trigger ordinary brain cells to suddenly join together. There also has to be something controlling the size of these assemblies. Greenfield believes this is a chemical process that relies not just on neurotransmitters in the brain but on the immune and hormonal systems of the body. It is also dependent on sensory inputs and feedback from the limbs and organs. So without the body, or the right chemicals to replace it, consciousness would not function properly. Bad news for any cryonics patients who've only had their head frozen.

WHY AM I?

Chemical, physical, electrical, quantum, emergent – consciousness could turn out to be any, all or none of these theories. There are plenty more. But why do we have it in the first place? Have we evolved consciousness or is it an inevitable side-effect of the evolution of our brain?

Evolution is driven by natural selection. The organisms best adapted to particular circumstances will be the ones that survive and get to breed. Desirable genetic traits are passed on to subsequent generations. Human beings are a result of more than 3.5 billion years of genetic changes, many virtually insignificant, others of major importance. Along the way some genetic mutations have had unexpected side-effects. So is consciousness a desirable characteristic that evolved through natural selection to make us better? You won't be surprised to hear that no one knows the answer but there are good arguments both ways.

GOOD NEWS FOR FIDO

One of the most respected voices on the evolution of consciousness is Cambridge academic Horace Barlow. He proposes that consciousness is what makes us able to interact socially with other humans and has been crucial for our development and survival. We stand out above other forms of life because of our social behaviour and our social (and antisocial) interactions with each other. Whenever we're conscious, he argues, we are addressing our thoughts to someone else.

Computer scientist Hans Moravec puts this another way. He thinks that as human language evolved, we told stories to each other about external and internal events (what we were thinking). Early on the 'storytelling mechanism was turned back on the teller', so it became a commentary on life.

Professor Barlow suggests that this introspection has given us the ability to understand what is going on in the minds of others. Comprehending other people's point of view is crucial to human social interaction. A person who is autistic lacks this ability to a greater or lesser degree. So the theory not only explains why consciousness might give us an evolutionary advantage but may account for the origins of autism.

The 'global workspace theory' developed by Bernard Baars at the Wright Institute in California, suggests a link between consciousness and the formation of memories. The global workspace refers to all the nerve signals, or qualia, clamouring for the brain's attention. With the driving example, the global workspace would be the road, your partner, the radio and perhaps your bladder. The most dominant of these becomes the consciousness for that period of time, and it's this dominant thought that the brain turns into memories.

If consciousness *has* evolved it would suggest that there's a sliding scale of consciousness and that it's not unique to humans. Other animals possess it too, just not as much of it – which would make dogs, dolphins, monkeys and marmots (well, maybe not marmots) conscious creatures. It certainly explains Lassie's ability to rescue little Jimmy from a mineshaft, but not perhaps Snoopy's delusional behaviour. Skippy, being a bush kangaroo, wouldn't have a great deal of brain so is, sadly, unlikely to be particularly conscious. More likely to be run down by an Ozzie redneck and served up on a barbecue.

But we digress.

Consciousness could just be an unintended consequence of evolution. Take dreams for instance. They might not serve any useful purpose in themselves but be a side-effect of the body's rejuvenation process. We have evolved the physical processes that go on in our sleep but not necessarily the dreams that accompany them. The same might be true of consciousness. The brain could be doing useful things like processing memories and consciousness is created as a result.

FOR EVER. HOW LONG IS THAT EXACTLY?

It's probably fair to say that the majority of scientists reckon consciousness will turn out to be something physical – a process generated purely within the grey matter of our brains and not some extra-sensory attribute that exists on a higher plane. If this proves to be the case, many philosophers and mystics will need to have a serious rethink about their careers – and probably therapy.

Assuming that consciousness is locked into our hard-wiring, whether at a cellular, molecular or quantum level, it should, in theory at least, be possible to replace the brain with an electronic copy. It might not even have to be a mass of specialised circuitry – maybe something an ordinary computer could cope with. With the brain as software on a PC, the possibilities become almost limitless. More on this later. It should also be possible to revive the consciousness after an extended period, which is good news for anyone cryogenically frozen. After all, consciousness disappears under general anaesthetic, only to return on waking.

Our goal from the beginning has been to live for ever. But there's a lot that can happen between now and then. For ever in this instance doesn't mean immortality. Science might be able to bring you close, but it can never completely guarantee it. If you die, you die. Downloading a back-up brain on to a computer might preserve your mind – but then what happens if the computer breaks down or someone accidentally wipes the disc or forgets to back it up? It would certainly make a 'fatal exception error' more literal.

Immortality is probably not that desirable anyway. Even the most self-confident, lively, socially stimulated individual would get a bit jaded after a few millennia. Even if cryonics offers the chance to sleep for a few thousand years, for ever is still a very long time. Immortality is often portrayed in fiction as something of a curse. *The Hitchhikers Guide to the Galaxy* has a bored space alien ('Wowbagger the Infinitely Prolonged,' since you ask) travelling the universe insulting everyone in it, just for something to do. An episode of *Doctor Who* ('The Five Doctors') has the villain seeking immortality only to end up encased for ever in stone as a conscious statue. (That's 90 minutes summed up in one sentence – other things happen along the way, mostly in a quarry.)

Impact Earth!

The idea that all life on Earth will be wiped out by an asteroid used to be the sole preserve of B movie producers scraping the barrel of pulp science fiction. This is no longer the case. Which is a relief to the scientists who have been warning for years of the very real dangers asteroids pose to our planet.

Nowadays there are international conferences on what are termed 'near earth objects' as well as government committees and scientific projects devoted to the study of asteroids. There's even an index of risk called the Torino Impact Hazard Scale that assigns a number to the chances of a particular asteroid colliding with the Earth. Zero means relax. Ten means it's too late. The system was devised by Richard Binzel, an astronomer at MIT, and is named after the Italian city where the Astronomical Union adopted it. We can only be grateful that they didn't hold the meeting in Ramsbottom. The Torino Scale is designed, as much as anything, to put all our minds at rest. Otherwise, scientists felt, people were getting unduly alarmed every time astronomers noticed anything that might have the remotest chance of hitting us.

The Earth is surrounded by cosmic debris. Tiny fragments hurtle towards the planet every day. The largest chunks of rock and the stuff of sci-fi nightmares are the asteroids. The smaller bits are called meteoroids. Most of these meteoroids become meteors, burning up in the atmosphere as shooting stars. Others, usually a few centimetres across and known as meteorites, occasionally make it to the ground. The chances of rocks hitting the Earth are quite high but the likelihood of the overwhelming majority of them doing any damage is extremely low.

However, there is a small chance that a very large rock might strike the Earth, and the resulting global impact would prove disastrous. The last one to cause serious damage exploded above a place called Tunguska in Siberia in 1908. Despite only being 60 metres in diameter, it completely destroyed 2,000 square kilometres of forest, having a power equivalent to that of several hydrogen bombs. Fortunately, the area was sparsely populated, or it could have been a lot worse. London, New York or Tokyo might just as easily have been hit – the odds are exactly the same.

The global consequences of an asteroid impact are almost too colossal to imagine. A 1-kilometre-wide lump of rock hitting the Earth would not only destroy everything in its path but would also throw tremendous amounts of dust into the atmosphere, altering the planet's climate. If it hit the ocean it would trigger tidal waves and flooding. Anything bigger and we are really in trouble.

It is widely accepted that the effects of an asteroid impact wiped out the dinosaurs. There is now an (albeit limited) international effort to ensure that next time we will at least see it coming, and a British-based organisation, Spaceguard, leads the way. This raises the question, if astronomers spot a lump of rock hurtling towards us, what do we do about it? Scientists then have to fall back on science-fiction scenarios such as those of *Armageddon* and *Deep Impact*, in which daring astronauts are sent to knock the rock off course. Which is precisely what is planned. If an Earth-bound asteroid is spotted early enough, it will only take a small nudge to deflect it. Too late and only Bruce Willis can save us.

The great thing about living for ever, as opposed to immortality, is that you usually get to decide when to pull the plug. But even if you want to live for a very long time, it's probably worth thinking about how long that could possibly be. After all, the Earth isn't going to be around for ever and if the majority of physicists are right, neither is the universe.

THE END OF THE WORLD

Assuming that nuclear or biological war, global warming, influenza or space invaders don't kill us first, there will come a time when the Earth can no longer support life. The planet we are on is, after all, just a tiny speck of dust compared to the bigger cosmic picture. At any moment the Earth is spinning on its axis, travelling around the Sun, which is travelling through the galaxy, which is travelling through the universe. Which means we're all flying through space at around 3 million kilometres per hour.

The Earth is part of a cosmic neighbourhood. There's the chance we could be hit by an asteroid. But even if we're not, we're still doomed. All terrestrial life on Earth ultimately depends on the Sun, or rather the Sun's energy, and that's not going to last for ever. When the Sun goes, it's going to take us with it. So if you were contemplating spending eternity on this small green and blue oasis, you may have to revise you plans.

The Sun, like any other star, is a giant atomic reactor. Inside, atoms of hydrogen are being constantly fused together to produce helium. The massive amounts of energy produced are given off as heat and light. All stars change as they get older, a process known as solar evolution. Eventually the Sun will start to run out of hydrogen fuel and begin to collapse. As pressure and temperature increase, the star will switch to burning its helium. The dying Sun will become larger and redder – a red giant.

You don't really want to get in the way of a giant, particularly a red one. As it expands it will swallow the nearest planets, first Mercury, then Venus. Until very recently the predictions were that the fiery ball would also eat up the Earth, but scientists at the University of Sussex have now calculated that our planet will escape destruction. As the Sun burns up its hydrogen and then helium to produce energy, although its size will get larger, its mass will get smaller. This means its gravity pull will also decrease, so the Earth will orbit further away from the now giant star.

So what sort of timeframe are we talking here? The Sun's got at least 7.5 billion years left. Dr Robert Smith from the University of Sussex reckons this gives the Earth around 5.7 billion years before it can no longer sustain life. Others are not quite so optimistic. Professor James Kasting from Pennsylvania State University has also done some calculations. He believes the Earth's oceans will all have boiled away in just over 1 billion years.

If you could live for ever, in reality this would mean somewhere between 1 and 5 billion years on Earth.

But we don't want to give you that…

To boldly go

Whichever way the Earth is eventually destroyed, we need to think about getting off if we want to survive. The answer, of course, is to build a starship

and seek out strange new worlds, new life and civilisations. So where will our starship go? It's all very well setting off on a fabulous interstellar journey, but there's not really much point if you haven't got any idea of where you're going. Ideally, we'd be bound for an Earth-like planet, with breathable air, fresh water and some plant or animal life to munch on. Science-fiction series like *Star Trek* suggest that the universe is littered with planets like this. You don't see the crew teleported onto terrain where they're instantly asphyxiated by poisonous fumes (apart from that unnamed crewman in the red shirt).

The chances are there *are* other planets elsewhere in the universe that can support life but astronomers don't know how many. It's generally assumed that for life to exist a planet has to have liquid water. Certainly to sustain any *human* life it will need water, or at least the means to make it. The Earth occupies a region of space known as the habitable, or 'Goldilocks', zone. Our planet is at just such a distance from the Sun that it's neither so hot that any water is vaporised nor so cold that it all freezes.

Even with the most powerful telescopes, astronomers cannot actually see any other planets beyond our immediate solar system. It doesn't mean that they aren't there, just that they are hidden, swamped by starlight. Their existence can only be deduced by looking at wobbles in the movements of stars caused by these 'invisible' planets' gravitation. So far scientists have found more than 100 'exoplanets', all of them gas giants like Jupiter. But as there are around 200 billion stars in our galaxy, they are almost certain to find some smaller, solid lumps of rock. How many are like Earth is a different question altogether, and it's one on which astronomers are divided.

Some computer models, such as one developed by Professor Barrie Jones at the Open University in Milton Keynes, calculate that there are as many as 1 billion Earth-like planets. However, with their 'rare earth hypothesis', geologist Peter Ward and astronomer Don Brownlee from the University of Washington in Seattle suggest that the circumstances that make Earth suitable for complex life are extremely rare. Key to their argument is the moon, which stabilises the tilt of our planet's axis. Without it there would have been enormous changes in climate holding back the evolution of higher life forms like us. That's not to say primitive life couldn't have evolved. Many astronomers are now of the opinion that life in the universe is widespread but intelligent life is not.

So, as part of our quest for eternal life, let's assume that we managed to hitch a lift on a starship. Maybe our consciousness is locked into the ship's computer software or maybe we're cryogenically frozen. Assuming also that there are other Earth-like planets, we can land on them and, in theory, survive. If there's life, maybe we can eat it. As long as there's water and a Sun, we have everything we need. The planet's atmosphere might not be quite right, perhaps failing to shield us from harmful radiation, but we could certainly survive. Until this new planet is wiped out. A few more space hops and maybe we've added on another couple of billion years to our existence.

But we don't want to give you that...

THE END

The whole point is to live for ever, to continue to exist whatever the fate of Earth or whatever planet we end up on. But 'for ever' rather depends on the fate of the universe. And once the universe goes, that really is it. So how long is that going to last?

In 1929, astronomer Edwin Hubble was able to show that the universe was expanding. He was the first to discover galaxies other than our own and calculated that these were getting further and further apart. This rate of expansion has been named the 'Hubble constant'. Scientists are now fairly convinced that this expansion began with a Big Bang. How it ends rather depends on the nature of the expanding universe. There are three options: it will go on expanding for ever, it will reverse its expansion or it will slow down and stop.

The first option, a universe of infinite expansion, is said to be 'open'. With this theory, once the stars have all died and the black holes evaporated, the universe will consist of decaying particles dissipated throughout a vast black void of cold nothingness. There will be no interactions and no chance of the universe being reborn.

The second theoretical model, in which the expansion reverses, imagines the universe as a sphere. Known as the 'closed universe', this model features the 'big crunch'. Whether the closed model is correct rather depends on how much material there is out there. Because of its mass, the Earth has enough gravity to stop us drifting off into space whenever we leave the house. Likewise, if the universe has enough mass and therefore enough gravity to pull galaxies back together, then the universe will start to collapse in on itself. In this scenario, planets and stars are torn apart as galaxies merge. As everything is pushed closer together, oceans boil away, planets evaporate and atoms cease to exist. Eventually, all the forces of matter unite. Scientists haven't a clue what happens next.

Compared to that, the final theory, in which the universe's expansion slows and stops, is a bit of an anticlimax. This is the currently fashionable idea of the 'flat universe', in which a balance is reached between infinite expansion and dramatic contraction. The flat universe theory relies on there being just the right amount of mass. But at the moment astrophysicists haven't been able to find it all, which is why you'll hear them muttering about 'dark matter'. This is the 'missing' matter that should be there to prove the flat universe theory right. There are experiments going on all over the world to prove that dark matter exists. If it turns out not to, the flat universe theory won't either.

So how long could we exist in an expanding, flat or crunching universe? The universe has been around for somewhere between 10 and 15 billion years, which on a cosmic scale is just a fraction of what's left to come. In their book, *The Five Ages of the Universe*, American astrophysicists Fred Adams and Greg Laughlin estimate that if the universe is expanding or flat it has got around 10^{100} years left. That's 1 followed by 100 zeros. More than all the grains of sand on all the beaches in the world, more than all the stars in the observable universe, more than anything we can possibly imagine. A long time. If, on the other hand, it starts collapsing in on itself towards a big crunch, the minimum amount of time left is a mere 50 billion years.

The infinite darkness of an expanding or flat universe would be a fairly miserable place to exist in even if it were possible. The final years of a big crunch would certainly be a more spectacular way to go, because there is absolutely no way anything could survive.

REST IN PEACE

The solution to living for ever, of course, is to build yourself a time machine and hop continuously backwards and forwards through space and time. While you're at it, you could start employing some of the other technology we've suggested. However you plan to spend eternity, you'll have to do something about your body every 70 years or so, replacing it with an electronic model or even cloning a new body and downloading the contents of your mind into it.

The alternative is to ditch your body completely, which might at first seem an appealing possibility. Particularly as you get older and things stop working properly. But what would it be like to be a mind inside a machine? A consciousness on a compact disc? At first it might be fun, flitting around the worldwide web peering through all those webcams, accessing the intimate secrets of nations, companies and individuals. But it probably wouldn't be long before you started to feel a bit isolated, cut off from the rest of the world. OK, so you could maybe interact with people through chat rooms or other disembodied consciousnesses, but you couldn't interact with real people in the real world.

Perhaps you wouldn't want to. You might choose to spend eternity in a series of virtual worlds, being a despot or a hero to virtual communities. If the virtual world was realistic enough, you could do want you wanted, when you wanted. Assuming you had control. But in a computer that's not necessarily the case. Your consciousness might be hijacked, traded or wiped out by a computer virus. Perhaps a black market in virtual consciousnesses would evolve. Your cyber dream would become someone else's cyber hell.

Quite aside from the practicalities, a few billion years of life probably wouldn't be that much fun. Once you'd got all your fantasies out of the way, it would most likely start to become somewhat mundane. Friends and partners would come and go, civilisations rise and fall, planets be born and die. There's a lot to be said for living a bit longer through gene therapy, nanotechnology or cybernetics and for making that life as fun as possible, but as for ever is a long time, we suggest that living for it is probably overrated.

You can easily achieve that feeling of for ever within an ordinary lifetime. There's a joke about the American State of Iowa, although you could just as easily substitute Lincolnshire or most of Belgium. A woman goes to her doctor who tells her she only has eight months to live. 'What can I do?' she asks. He tells her to marry an economist and move to Iowa. 'Will that extend my life?' 'No,' he says, 'but it'll sure seem longer.'

Whether you're religious or not, dying is probably not as bad as you think.

Suggested Reading

We will spare you the scientific papers that are referred to throughout *How to Clone the Perfect Blonde* and instead point you in the direction of genuinely enlightening science books. There are many others of course but if you want to follow any of the chapter subjects in more depth we think these are among the best reads.

Clone, Gina Kolata (Penguin). A marvellously thorough account of the science that led up to Dolly the sheep.

Second Creation, Ian Wilmut, Keith Campbell and Colin Tudge (Headline).

Genome, Matt Ridley (Fourth Estate). Or indeed anything by Ridley.

A passion for DNA, James D. Watson (Oxford University Press). A series of essays by the co-discoverer of the double helix.

Robot, Rodney A. Brooks (Penguin). From the Director of the Artificial Intelligence Laboratory at MIT. A very readable account from the cutting edge of AI.

Emergence, Steven Johnson (Penguin). Excellent insight into this fascinating branch of science.

Computer, Campbell-Kelly, William Aspray (Basic Books). One of the few books on the history of computers. If you're interested in this sort of thing, another of the more entertaining is *Fire in the Valley,* Freiberger and Swaine (McGraw-Hill).

The Physics of Star Trek, Lawrence Krauss (Flamingo).

Time Travel in Einstein's Universe, J. Richard Gott (Phoenix). A little complicated in places but is generally a good read.

The Future of Spacetime, Stephen Hawking, Kip Thorne, Igor Novikov, Timothy Ferris and Alan Lightman (Norton). As you can guess from the number of authors, this is a collection of essays. An entertaining account of black holes, gravitational waves and time travel.

How to Build a Time Machine, Paul Davies (Penguin).

The Fabric of Reality, David Deutsch (Penguin).

In Search of Schrödinger's Cat, John Gribbin (Corgi). In fact, any science book by John Gribbin because he can explain most things simply.

Black Holes and Time Warps, Kip Thorne (Papermac). Extremely comprehensive but some sections are, in our opinion, far too difficult for the average reader. So provided you are selective you'll be ok.

Consciousness, Rita Carter (Weidenfeld and Nicolson). There aren't many popular science books on the mind but this is among the best.

Robot, Hans Moravec (Oxford University Press). Either an insight into the future or the rantings of a madman. You decide.

The Five Ages of the Universe, Fred Adams and Greg Laughlin (Simon and Schuster). Heavy going in places but still an excellent guidebook to the long-term future of the universe.

There are a few science-fiction books that can't pass without mention: H. G. Wells' *The Time Machine, About Time* by Jack Finney (Simon and Schuster); *I, Robot* by Isaac Asimov, *The Hitchiker's Guide* series by Douglas Adams and *Do Androids Dream of Electric Sheep* (also known as *Bladerunner*) by Philip K. Dick. Of the films we've mentioned, for the sake of your own sanity don't bother with *Fantastic Voyage* (Twentieth Century Fox), certainly not sober. It was laughable 30 years ago and doesn't improve with age.

Index